Digital Design with CPLD Applications and VHDL: A Lab Manual

Robert K. Dueck
Red River Community College

DELMAR
THOMSON LEARNING™

Australia Canada Mexico Singapore Spain United Kingdom United States

Digital Design with CPLD Applications and VHDL:
A Lab Manual
by Robert K. Dueck

Business Unit Director:
Alar Elken

Executive Editor:
Sandy Clark

Senior Acquisitions Editor:
Gregory L. Clayton

Developmental Editor:
Michelle Ruelos Cannistraci

Editorial Assistant:
Jennifer Thompson

Executive Marketing Manager:
Maura Theriault

Channel Manager:
Mona Caron

Marketing Coordinator:
Paula Collins

Executive Production Manager:
Mary Ellen Black

Senior Project Editor:
Christopher Chien

Production Manager:
Larry Main

Art/Design Coordinator:
David Arsenault

Printed in Canada
4 5 XXX 05 04 03

For more information contact
Delmar, 3 Columbia Circle,
PO Box 15015,
Albany, NY 12212-5015.
Or find us on the World Wide Web at
http://www.delmar.com

ISBN 0-7668-1161-1

NOTICE TO THE READER

Contents

Preface

Intended Audience

This series of labs is intended for use in a digital systems or digital design sequence, as part of a Computer/Electronics Engineering Technology (CET/EET) program. The labs are based on material in *Digital Design with CPLD Applications and VHDL* (Robert K. Dueck, Delmar Publishers, © 2001).

Programmable Logic as a Vehicle for Teaching Digital Design

Historically, digital logic or digital design courses at the EET level have focussed on using fixed-function TTL and CMOS integrated circuits as the vehicle for teaching principles of logic design. However, the digital design field has changed; more and more, digital designs are being implemented in Programmable Logic Devices (PLDs), rendering many of the popular fixed-function devices obsolete.

This lab manual, and the textbook it accompanies, address this trend by focussing primarily on PLDs as a vehicle for teaching digital logic. The labs use the Student Edition of Altera's MAX+PLUS II PLD design and programming software. The exercises in this manual are configured for use with either of two hardware platforms: the Altera UP-1 University Program Laboratory Design Package or the RSR PLDT-2 Programmable Logic Trainer.

Both boards contain an Altera EPM7128S Complex Programmable Logic Device (CPLD), a Sum-of-Products (SOP) device with 128 macrocells, based on EEPROM technology. This chip is in-system programmable and thus can be programmed and erased multiple times, via a cable from a PC parallel port, without removing it from the board. In addition to the CPLD, both boards include a number of standard input and output devices, such as switches, LEDs, and seven-segment numerical displays.

The MAX+PLUS II software is bundled with *Digital Design with CPLD Applications and VHDL* textbook. Each installed copy must be activated by a license file available, free of charge, by e-mail from the Altera web site (**www.altera.com**).

The Altera UP-1 circuit board is available from Altera's University Program for sale to students ($149.00, as of September, 2000) or can be requested by educational institutions, either for purchase or on a donation basis for those institutions that are members of the Altera University Program. Institutions can also request donations of full-version software and CPLD chips.

The RSR PLDT-2 board, available from RSR Electronics/Electronix Express (**http://www.elexp.com**), can be purchased for $89.00 in quantities of 10 or more (as of June, 2000). This board has some features, such as jumpered inputs and outputs, debounced switches, clearly numbered pin connections, active-HIGH LEDs and lots of wiring room around the chip, that make it more user-friendly than the Altera UP-1 board.

Traditional SSI Labs

I have found it useful in my own teaching to start the digital lab sequence with a few exercises based on traditional SSI components. Hands-on wiring of devices teaches something about circuit construction that you don't get in software-based design entry. Labs 1, 2, and 3 give some opportunity for circuit breadboarding, where students get practice finding chip pin numbers, wiring discrete components, and connecting (or forgetting to connect) power supplies to circuits. These labs give the student useful reinforcement of basic digital principles in the first few weeks of the course before the CPLD-based material is introduced.

The experience of breadboarding is later combined with the software-based design entry in Labs 16 and 17.

CPLD Labs

Labs 4 to 17 make use of the various design entry and simulation features of MAX+PLUS II, including schematic capture, text entry using VHDL (VHSIC Hardware Description Language; VHSIC = Very High Speed Integrated Circuit), and use of components from the Altera Library of Parameterized Modules (LPM). They are stepped in difficulty so that, in the initial exercises, a student is asked only to follow given procedures without making too many decisions and in later labs is given broad direction, requiring some independent decision-making. The principles taught in Labs 4 to 13 are brought together in a design project in Lab 14. Labs 15 to 17 are more advanced units (state machines, D/A conversion, A/D conversion) that stand on their own.

Class Time Requirements

The labs in this package are generally longer than digital labs in most of the current EET lab manuals available on the market. With the exception of Labs 1, 2 and 4, each of which can easily be completed in a single two-hour lab session, these labs are designed for about 4–6 hours each. The design project in Lab 14 requires about 10–12 hours.

In traditional TTL/breadboard labs, this would be problematic due to the time required for circuit connection and the inevitable task of troubleshooting the spaghetti forest that accompanies it. However, when programming CPLDs with MAX+PLUS II, the majority of design setup is done in software. This lends a flexibility and portability that has not been available until now. Students can work on their labs on their own time, at home or in a computer lab, requiring the hardware only for final design testing and demonstration.

Also, the CPLD pin assignments for Labs 4 to 15 are standard; it is possible to wire the switch/LED connections on the Altera UP-1 or RSR PLDT-2 board once and then forget it. ***Labs 4 to 15 can be demonstrated sequentially without ever rewiring the board.*** The standard wiring configuration for the CPLD board is given just before Lab 4. It is recommended that the user spend about half an hour wiring the CPLD board, then use the standard wiring for the laboratory exercises.

The wiring configuration is the same for both the Altera UP-1 board and the RSR PLDT-2 board. Thus, all labs can be run on either board without reassigning pin numbers. Due to some design differences in the boards (active-HIGH vs. active-LOW outputs and different on-board oscillator speeds), some adjustments might need to be made when transferring a design from one board to another. These differences are outlined in the individual lab exercises.

MAX+PLUS II Design Files

Two sets of design and programming files are available:

1. **For Instructors:** a full set of required Graphic Design Files (gdf), VHDL files (vhd) and Programmer Object Files (pof) for all labs, as well as the related project files for each design. These files are available from Delmar Thomson Learning as part of the *e-resource* package for *Digital Design with CPLD Applications and VHDL.* (ISBN: 0-7668-1252-9)
2. **For Students:** a limited number of design entry files required in some of the CPLD labs. These are generally for components that are needed to make a particular design work properly, but which the student is not expected to create at that point in the lab sequence. These files are included on the CD-ROM that accompanies this manual.

Acknowledgments

The author and Delmar Publishers would like to thank the following reviewers:

David Delker, Kansas State University, Salina, KS

Mike Miller, DeVry Institute of Technology, Phoenix, AZ

Bob Rowley, DeVry Institute of Technology, Phoenix, AZ

Carlo Sapijaszko, DeVry Institute of Technology, Orlando, FL

Gilbert Seah, DeVry Institute of Technology, Scarborough, Ontario

Lloyd Stallkamp, Montana State University, Havre, MT

Lab 1

DIP Integrated Circuits

Name SALMAN BASARIA Class ________ Date JAN 28 2005

Objectives Upon completion of this laboratory exercise, you should be able to:

- Describe the configuration of several basic logic gates in Dual In-Line Packages.
- Wire a logic gate IC on a prototyping breadboard.
- Obtain the truth tables of each gate to be tested.

Reference Dueck, Robert K., *Digital Design with CPLD Applications and VHDL*

Chapter 2: Logic Functions and Gates
2.1 Basic Logic Functions
2.2 Logic Switches and LED Indicators
2.3 Derived Logic Functions
2.6 Integrated Circuit Logic Gates

Equipment Required

+5-volt power supply
Breadboard
Wire strippers
#22 solid-core wire, as required

Components as follows:

Part No.	Qty.	Description
74LS00 or 74HC00	1	Quad 2-input NAND gate
74LS02 or 74HC02	1	Quad 2-input NOR gate
74LS04 or 74HC04	1	Hex Inverter
74LS08 or 74HC08	1	Quad 2-input AND gate
74LS32 or 74HC32	1	Quad 2-input OR gate
74LS86 or 74HC86	1	Quad 2-input XOR gate
	1	4PST or 8PST DIP Switch
	1	LED
	1	Resistor, 330 Ω, ¼ W
	2	Resistor, 10 kΩ, ¼ W

Experimental Notes

A common way to package logic gates is in a plastic or ceramic **Dual In-Line Package,** or **DIP,** which has two parallel rows of pins. The standard spacing between pins in one row is 0.1″ (or 100 mil). For packages having less than 28 pins, the spacing between rows is 0.3″ (or 300 mil). For larger packages, the rows are spaced by 0.6″ (or 600 mil).

The outline of a 14-pin DIP is shown in Figure 1.1. There is a notch on one end to show the orientation of the pins. When the IC is oriented as shown, pin 1 is at the bottom left corner and the pins number counterclockwise from that point.

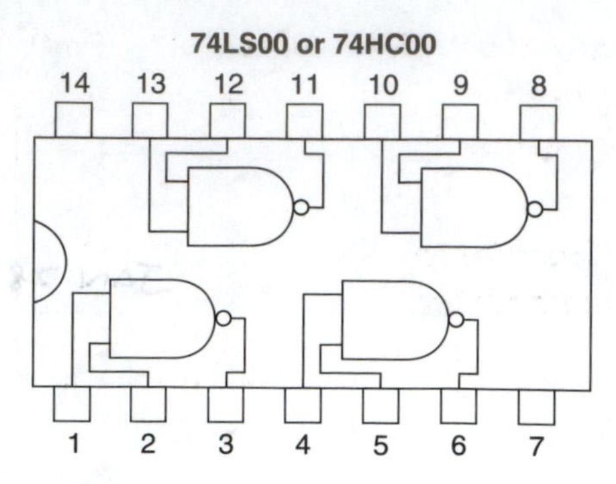

Figure 1.1 DIP IC (Quad 2-input NAND)

Figure 1.2 shows the internal diagrams of gates listed in the Equipment Required section. In addition to the gate inputs and outputs, there are two more connections to be made on each chip: the power and ground connections. V_{CC} (pin 14) must be connected to +5 volts and GND (pin 7) to ground to provide power supply connections. ***The gates won't work without these connections.*** Logic levels at the other pin inputs are derived from these power supply voltages by connecting them to +5 volts for logic 1 and ground for logic 0.

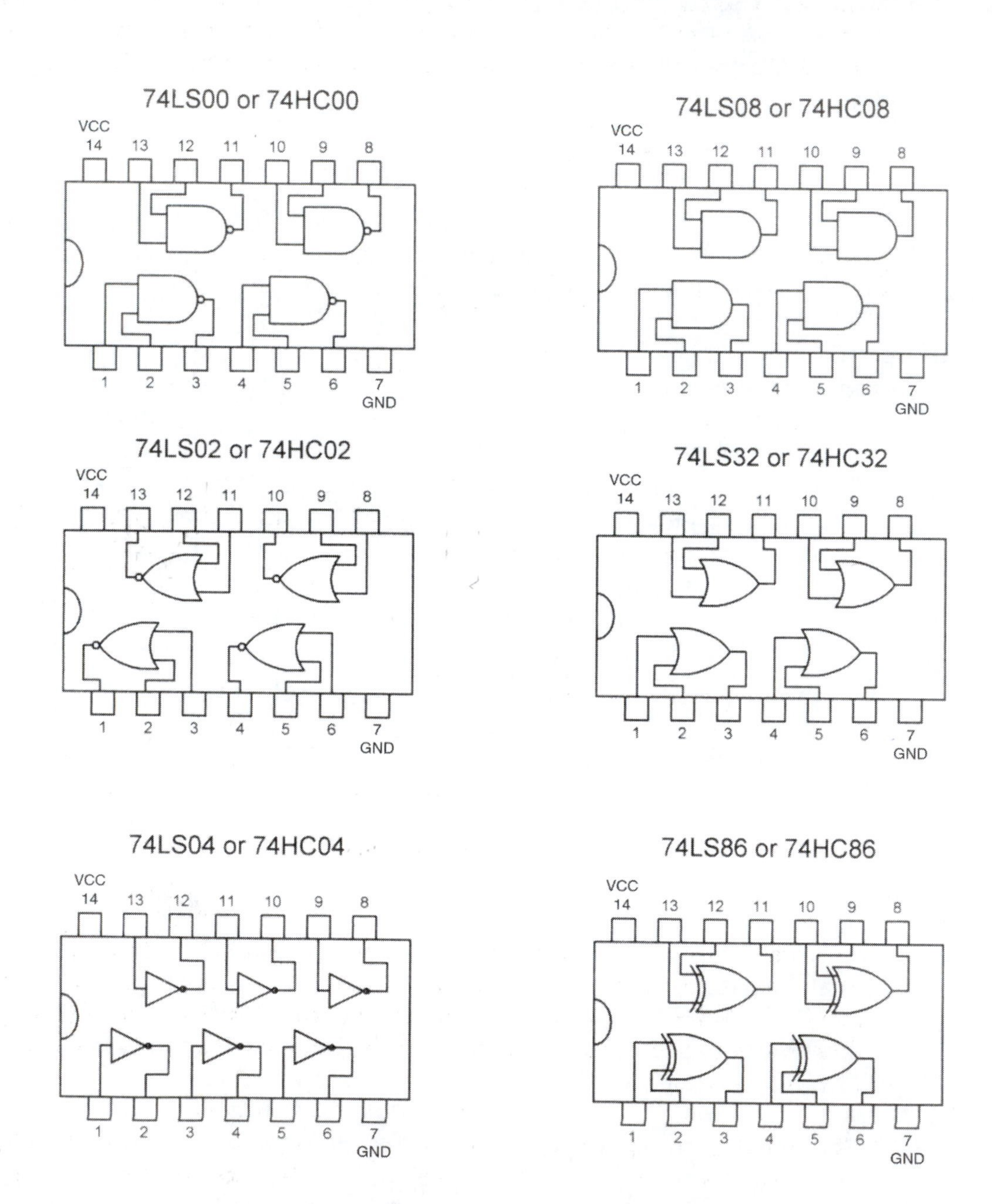

Figure 1.2 Pinouts of Some Basic Logic ICs

A **truth table** of a digital logic gate or circuit is a table showing the gate or circuit output for all possible combinations of inputs. These *must* be shown in standard binary order. Table 1.1 shows an example.

Note that the input combinations count up from 00 to 11 in binary. This is the standard order for all truth tables.

Table 1.1 **Truth Table**

A	*B*	*Y*
0	0	1
0	1	1
1	0	1
1	1	0

Procedure

1. Insert a 74LS00 or 74HC00 Quad 2-input NAND gate into the circuit breadboard. Connect pin 14 to V_{CC} (the + terminal of the +5-volt power supply) and pin 7 to ground (the – terminal of the power supply), as shown in Figure 1.3.

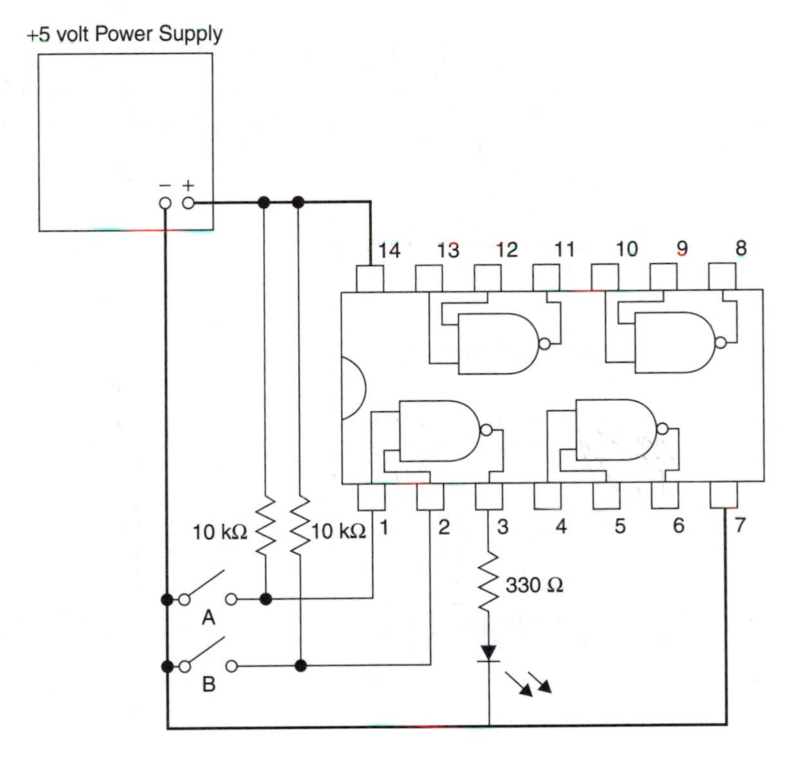

Figure 1.3 Testing a Gate in a DIP IC

2. If you are not using a digital trainer with logic switches and LED indicators, connect one end of a single-pole single-throw (SPST) switch to ground and the other end to pin 1 of the NAND IC. Connect the junction of the switch and pin 1 to V_{CC} through a 10-kΩ resistor. This represents input A of the gate. Make a similar connection to pin 2 of the IC for input B.

Note If you are using a digital logic trainer with logic switches and LED indicators, you may use those instead of the switches and LED shown in Figure 1.3. In this case, the resistors are not required. Simply connect pins 1 and 2 to the pins associated with the logic switches. If in doubt, ask your instructor.

3. Connect pin 3 of the NAND IC to an LED through a 330-Ω series resistor or to the pin for an LED indicator on a digital trainer. The cathode (indicated by a short lead or by a flat spot on the LED case) should go to ground.

4. Take the truth table of the gate by changing inputs *A* and *B* to make all possible combinations of input (*AB* = 00, 01, 10, 11) and writing down whether the LED at output 3 indicates logic 1 (ON) or logic 0 (OFF) for each input combination.

Table 1.2 NAND Truth Table

A	*B*	*Y*
0	0	1
0	1	1
1	0	1
1	1	0

5. Move the logic input connections to pins 4 and 5 and the lamp to pin 6 and repeat step 4. Also repeat with:

   logic switches at pins 9 and 10 and lamp at pin 8, and;

   logic switches at pins 12 and 13 and lamp at pin 11.

6. Repeat the above steps with the other gates listed in the parts list. (Note that the 74LS02/74HC02 and 74LS04/74HC04 gates require different connections than the rest of the gates. Consult Figure 1.2 for details.) Make a truth table for one gate in each IC. See Tables 1.3 through 1.7.

Table 1.3 AND Truth Table

A	*B*	*Y*
0	0	0
0	1	0
1	0	0
1	1	1

Table 1.4 NOR Truth Table

A	*B*	*Y*
0	0	1
0	1	0
1	0	0
1	1	0

Table 1.5 OR Truth Table

A	*B*	*Y*
0	0	0
0	1	1
1	0	1
1	1	1

Table 1.6 Inverter Truth Table

A	*Y*
0	1
1	0

Table 1.7 XOR Truth Table

A	*B*	*Y*
0	0	0
0	1	1
1	0	1
1	1	0

Instructor's Initials: ________

Assignment Questions

Do problems 2.4, 2.8, 2.10, 2.12, 2.14, 2.18, and 2.20 in the textbook and hand in the answers stapled to a cover sheet entitled Lab 1 Assignment Questions.

Lab 2

Pulsed Operation of Logic Gates

Name ______________________ Class __________ Date ____________

Objectives Upon completion of this laboratory exercise, you should be able to:

- Determine how basic logic gates can be used to pass (enable) or block (inhibit) time-varying digital signals by examining the gate truth tables.
- Monitor the pulsed behavior of logic gates with LEDs and with an oscilloscope.

Reference Dueck, Robert K., *Digital Design with CPLD Applications and VHDL*

Chapter 2: Logic Functions and Gates
2.2 Logic Switches and LED Indicators
2.5 Enable and Inhibit Properties of Logic Gates

Equipment Required

+5-volt power supply
Breadboard
Wire strippers
#22 solid-core wire, as required

Components as follows:

Part No.	Qty.	Description
74LS00 or 74HC00	1	Quad 2-input NAND gate
74LS02 or 74HC02	1	Quad 2-input NOR gate
74LS08 or 74HC08	1	Quad 2-input AND gate
74LS32 or 74HC32	1	Quad 2-input OR gate
74LS86 or 74HC86	1	Quad 2-input XOR gate
NE555 or equivalent	1	555 Timer
	2	LED, red
	2	Resistor, 330 Ω, ¼ W
	1	Resistor, 10 kΩ, ¼ W
	1	Resistor, 100 kΩ, ¼ W
	1	Capacitor, 820 pF
	1	Capacitor, 0.01 μF
	1	Capacitor, 4.7 μF

Experimental Notes

Any logic gate can be used as a switch to pass or block a time-varying waveform. When the waveform is passed by the gate, we say that the gate is enabled; when the waveform is blocked, the gate is inhibited. Each type of logic gate has a particular set of enable/inhibit characteristics: the gate may be enabled by a logic 1 or a logic 0; the time-varying input might pass through the gate in true or complement form.

We can predict the gate characteristics by examining the truth table of the gate. For example, the truth table of an AND gate is shown in Table 2.1.

A pulse waveform is just a sequence of logic 0s and 1s. If we hold input *A* at a constant level and vary input *B* between logic 0 and logic 1, we will get a waveform similar to the first half of the timing diagram in Figure 2.1. If we then change *A* to its opposite level, we will get a waveform like the one in the second half of Figure 2.1.

Table 2.1 AND Truth Table Showing Enable/Inhibit Properties

A	*B*	*Y*	
0	0	0	Y = 0
0	1	0	(Inhibit)
1	0	0	Y = 1
1	1	1	(Enable)

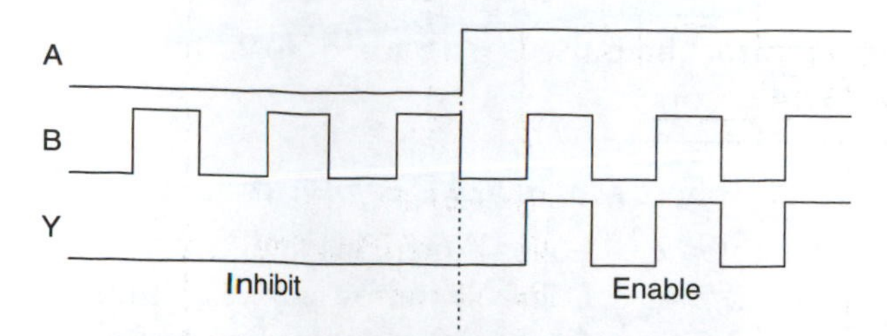

Figure 2.1 Enable/Inhibit Waveforms

If input A is 0 and B is pulsing, the gate output is always 0, since an AND gate requires all inputs to be HIGH to make the output HIGH. This never happens when *A* = 0. The first two lines of the truth table in Table 2.1 show this condition. If *A* is 1 and *B* is pulsing, the gate output is the same as *B*, as shown by the last two lines of the truth table.

We can designate the *A* input as the "control" input, since it controls whether the waveform at *B* will pass through the gate or not. Input *B* is designated as the "signal" input. Our examination of the AND gate truth table tells us that an AND gate is enabled by a 1 at its control input and that the signal is passed through in true (noninverted) form when the gate is enabled. Other logic gates have similar characteristics, as predicted by examination of their truth tables.

Procedure

1. Connect the circuit shown in Figure 2.2, using a 74LS08 or 74HC08 Quadruple 2-input AND gate. The connections labeled Fast and Slow are wire jumpers that select between periodic waveforms having frequencies of approximately 8.4 kHz and 1.5 Hz, respectively. The former is suitable for viewing on an oscilloscope, while the latter is slow enough to observe directly on the LEDs. The logic level at the gate's control input is also selected by a wire jumper.

   If you are using a digital trainer, the components in the box labeled **Pulse Source** are not necessary. In this case replace the pulse source with a connection to a clock generator or TTL-level pulse signal provided on the trainer. Use about 1 Hz for the slow speed and 10 kHz for fast. The wire jumper for logic 0 and 1 at the gate input can be replaced by a logic switch, or a DIP switch with a 10-kΩ pull-up resistor, as shown in Lab 1.

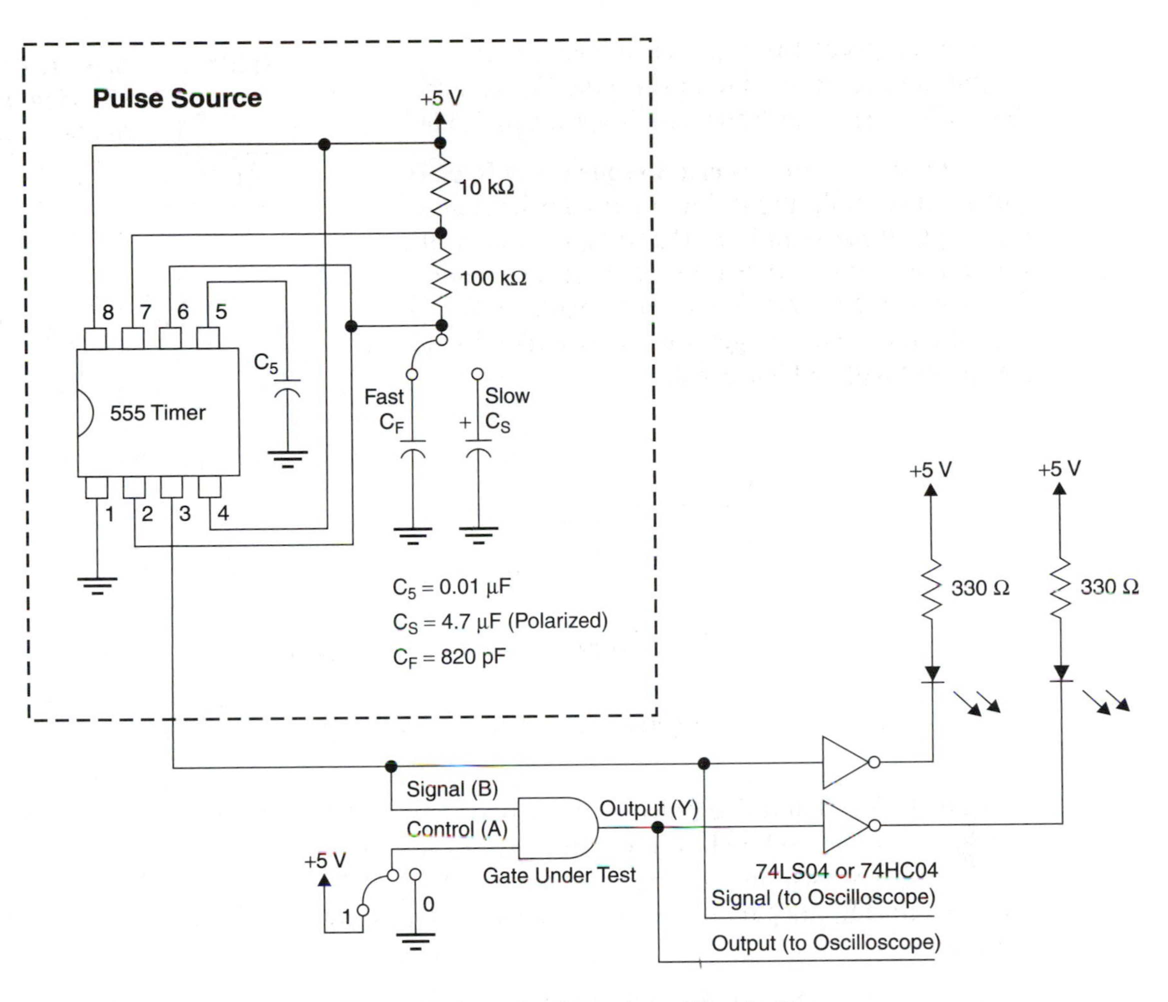

Figure 2.2 Pulsed Testing of Logic Gates

2. Set the Control input to 0 and the pulse source to **Slow**. Monitor the signal input and the gate output on the LEDs. How do the B and Y LEDs relate to each other when:

   a. $A = 0$? ____

   b. $A = 1$? ____

   (Is the Y LED always ON, always OFF, flashing with B (in phase), or flashing opposite to B (out of phase)?

Instructor's Initials: __________

3. Set the pulse source frequency to **Fast**. Monitor the signal input to the gate on channel B of the oscilloscope and the gate output on channel A. Trigger the oscilloscope on channel B. Draw waveforms similar to Figure 2.1 for the cases when $A = 0$ and $A = 1$.

Instructor's Initials: __________

4. Repeat procedures 1 through 3 for the following logic gates:

   a. 74LS00 or 74HC00 Quadruple 2-input NAND gate

   b. 74LS02 or 74HC02 Quadruple 2-input NOR gate

c. 74LS32 or 74HC32 Quadruple 2-input OR gate

d. 74LS86 or 74HC86 Quadruple 2-input XOR gate

Instructor's Initials: ________

Assignment Questions

1. Make a table relating control input, signal input, and output for all gates (e.g., for an AND gate, $Y = 0$ when $A = 0$; $Y = B$ when $A = 1$).
2. Hand in the results for this lab on a separate sheet. Draw a timing diagram for each gate, showing the relationship between control input, signal input, and gate output. When drawing the waveforms, use a ruler or drafting software.
3. Complete and hand in problems 2.26, 2.28, and 2.30 through 2.33 from the textbook.

Lab 3

Boolean Algebra

Name ______________________ Class __________ Date __________

Objectives Upon completion of this laboratory exercise, you should be able to:

- Draw the logic diagram of a combinational circuit from a Boolean expression.
- Take a truth table from a logic gate network and use it to derive a sum-of-products (SOP) Boolean expression for the network.
- Use Boolean algebra to simplify a logic gate network and to prove that two gate networks are equivalent.
- Use DeMorgan equivalent forms of logic gates to simplify the Boolean expression of a logic gate network.

Reference Dueck, Robert K., *Digital Design with CPLD Applications and VHDL*

Chapter 2: Logic Functions and Gates
2.4 DeMorgan's Theorems and Gate Equivalence
Chapter 3: Boolean Algebra and Combinational Logic

Equipment Required
+5-volt power supply
Breadboard
Wire strippers
#22 solid-core wire, as required
Components as follows:

Part No.	Qty.	Description
74LS00 or 74HC00	1	Quad 2-input NAND gate
74LS02 or 74HC02	1	Quad 2-input NOR gate
74LS04 or 74HC04	1	Hex Inverter
74LS08 or 74HC08	1	Quad 2-input AND gate
74LS10 or 74HC10	2	Triple 3-input NAND gate
	1	4PST or 8PST DIP Switch
	1	LED
	1	Resistor, 330 Ω, ¼ W
	4	Resistor, 10 kΩ, ¼ W

Experimental Notes

Any combinational logic circuit can be described by a Boolean expression written in sum-of-products (SOP) form. This form can be derived from a truth table by noting the lines on the truth table where the output has a value of 1 and writing a **product term** for each such line. Each product term consists of all input variables in either true or complement form. If an input is 0, it is written in complement form (with a bar); if the input is 1, it is written in true form (no bar).

For example, in Table 3.1 there are three product terms. These three terms are combined in an OR function to make a sum-of-products Boolean expression:

$$Y = \overline{A}\,\overline{B}\,\overline{C} + \overline{A}\,B\,C + A\,B\,\overline{C}$$

Figure 3.1 shows the logic diagram derived from the above Boolean expression. The circuit can be implemented by a combination of AND, OR, and NOT gates.

Table 3.1 Product Terms from a Truth Table

A	*B*	*C*	*Y*	
0	0	0	1	$\overline{A}\,\overline{B}\,\overline{C}$
0	0	1	0	
0	1	0	0	
0	1	1	1	$\overline{A}\,B\,C$
1	0	0	0	
1	0	1	0	
1	1	0	1	$A\,B\,\overline{C}$
1	1	1	0	

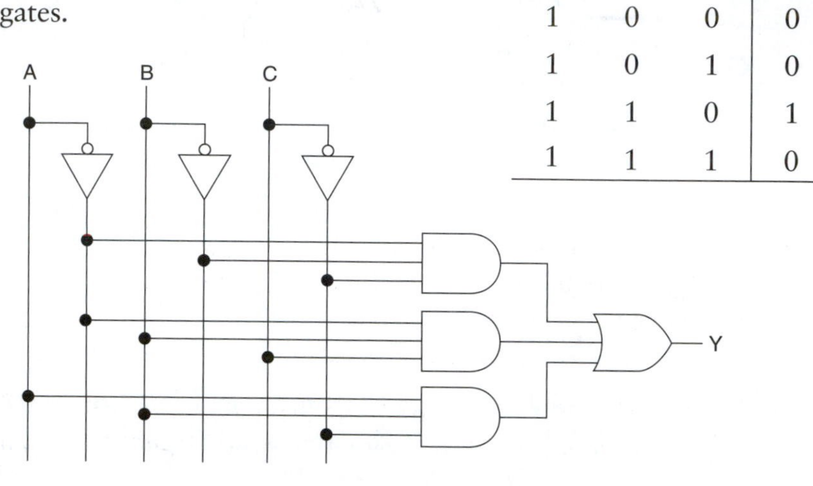

Figure 3.1 Sum-of-Products Network (AND-OR Construction)

An alternative configuration, shown in Figure 3.2, can be constructed using only NAND gates and inverters. The output gate is a NAND gate, shown in its DeMorgan equivalent form. The circuit can be derived from an AND-OR configuration by inverting all AND outputs and all OR inputs, as shown. In some cases, this can result in a more efficient circuit implementation (i.e., with fewer logic gate packages) than the AND-OR circuit. In this case, both circuits are equally efficient to build.

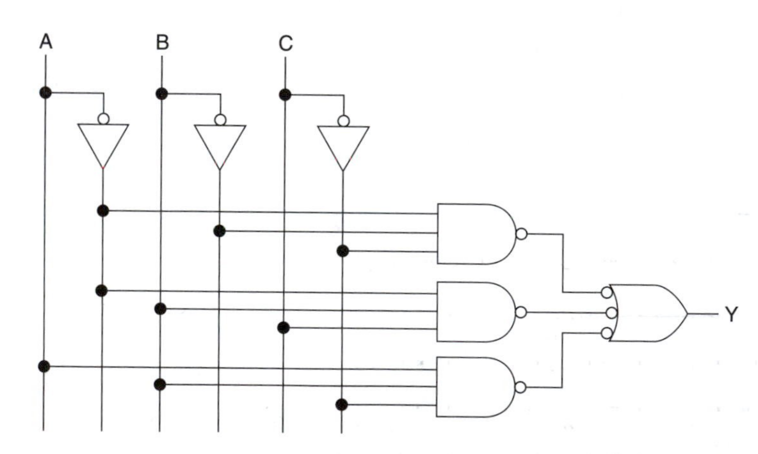

Figure 3.2 Sum-of-Products Network (NAND-NAND Construction)

Procedure

1. Draw the unsimplified logic diagram represented by the following Boolean expression:

$$Y = \overline{(A\,D + B\,\overline{D})\,C}$$

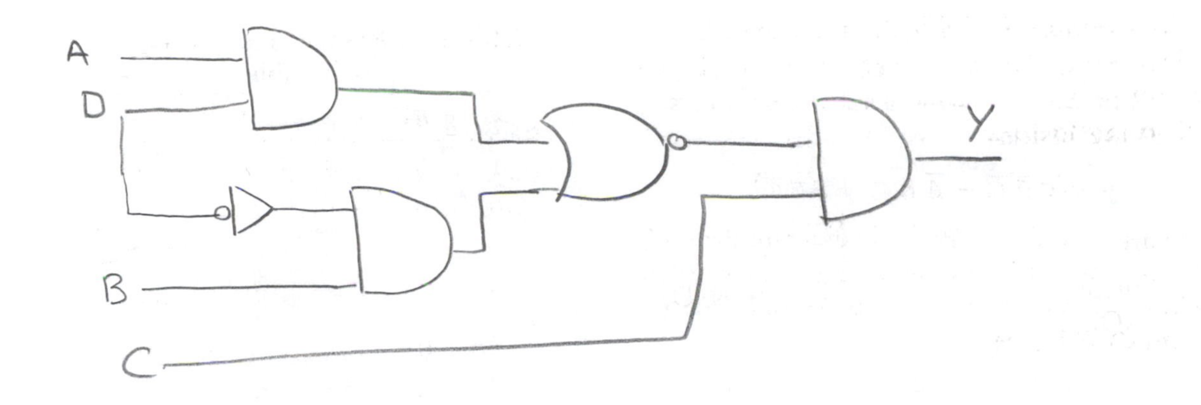

2. Connect the circuit drawn in procedure 1, using 74LS08 or 74HC08 AND gates, 74LS02 or 74HC02 NOR gates, and 74LS04 or 74HC04 inverters. Refer to Figure 3.4 at the end of this lab for the IC pinouts. Connect a logic switch to each input and an LED monitor to the output.

   Refer to Figure 1.3 (Lab 1) or Figure 2.2 (Lab 2) for configuration of logic switches and LED monitors. If you are using a digital trainer, use the logic switches and LED monitor on the trainer.

3. Construct the truth table of the circuit in procedure 2 by setting the input switches to all possible input combinations and noting the output value for each combination.

A	B	C	D	$Y = \overline{(AD + B\bar{D})}C$
0	0	0	0	0
0	0	0	1	0
0	0	1	0	1
0	0	1	1	1
0	1	0	0	0
0	1	0	1	0
0	1	1	0	0
0	1	1	1	1
1	0	0	0	0
1	0	0	1	0
1	0	1	0	1
1	1	0	0	0
1	1	0	1	0
1	1	1	0	0
1	1	1	1	0

Feb

4. Write the SOP expression derived from the truth table. Use Boolean algebra to simplify the expression as much as possible.

$$Y = \bar{A}\bar{B}C\bar{D} + \bar{A}\bar{B}CD + \bar{A}BCD + A\bar{B}C\bar{D}$$

$$= \bar{A}\bar{B}C + \bar{A}BCD + A\bar{B}C\bar{D}$$

$$= \bar{A}\bar{B}C + \bar{A}CD + A\bar{B}C\bar{D}$$

$$= \bar{B}C\bar{D} + \bar{A}CD$$

5. Draw the circuit described by the simplified Boolean expression in procedure 4. Connect the circuit using only 74LS10 or 74HC10 Triple 3-input NAND gates and 74LS04 or 74HC04 inverters.

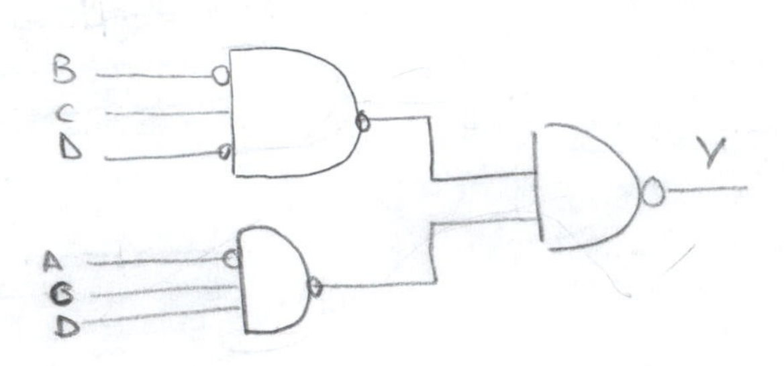

6. Take the truth table of the circuit drawn in procedure 5. Verify that it is the same as the table taken in procedure 3.

A	B	C	D	Y
0	0	0	0	
0	0	0	1	
0	0	1	0	
0	0	1	1	
0	1	0	0	
0	1	0	1	
0	1	1	0	
1	0	0	0	
1	0	0	1	
1	0	1	0	
1	0	1	1	

A	B	C	D	Y
1	1	0	0	
1	1	0	1	
1	1	1	0	
1	1	1	1	

4.M 2/18

7. Write the unsimplified Boolean expression for the logic diagram shown in Figure 3.3.

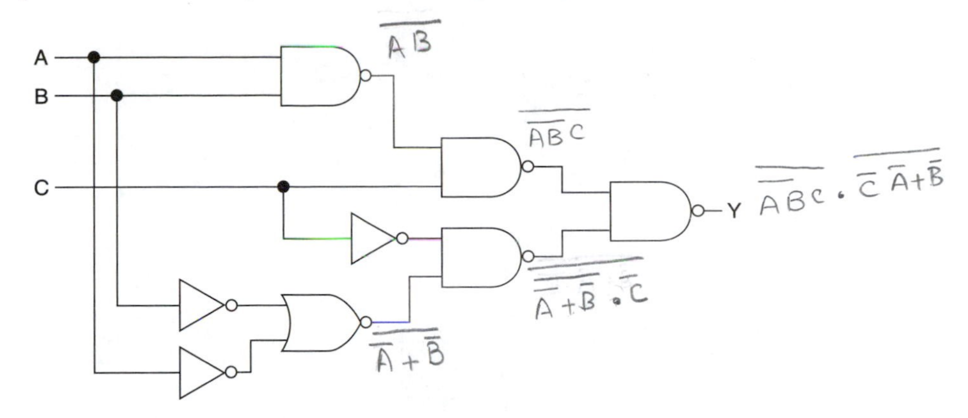

Figure 3.3 Logic Gate Network

8. Redraw the logic diagram in Figure 3.3, replacing some of the gates with their DeMorgan equivalent forms, so that each gate output with a bubble drives a gate input with a bubble and each gate output with no bubble drives a gate input with no bubble. Write the Boolean expression of the redrawn circuit. (For guidance, refer to examples 3.1 and 3.2 (p. 59), problem 3.3 (p. 107), and the answer to problem 3.3 (p. 778) in *Digital Design with CPLD Applications and VHDL.*)

$\overline{(A+B+\bar{C})\cdot(\bar{A}\bar{B}+C)}$

$\overline{AB}\cdot\bar{C}$

$\overline{(A+B+\bar{C})}+\overline{(\bar{A}\bar{B}+C)}$

$\bar{A}\bar{B}C + AB\bar{C}$

9. Construct the circuit of procedure 8 using the minimum number of IC packages. **(Note:** an inverter can be constructed from a NAND or NOR gate by connecting the gate inputs together.) Why? ______________________

____

____

Take the truth table of the circuit. Write the equivalent SOP expression of this circuit on the next page.

10. Use Boolean algebra to simplify the expression derived in procedure 8 to the simplest possible SOP form. Draw the logic diagram of the simplified expression. Use Boolean algebra to prove that this circuit is equivalent to the circuit constructed in procedure 9.

How many logic gate ICs are required to construct this circuit if AND-OR construction is used? ____________

Which ones? ____

11. Redraw the simplified SOP circuit of procedure 10 to use NAND-NAND construction, as in Figure 3.2. How many logic gate ICs are required to make this circuit?

____

Which ones? ____

Construct the circuit as a NAND-NAND circuit and take its truth table. Verify that this is the same as the truth table taken in procedure 9. Refer to Figure 3.4.

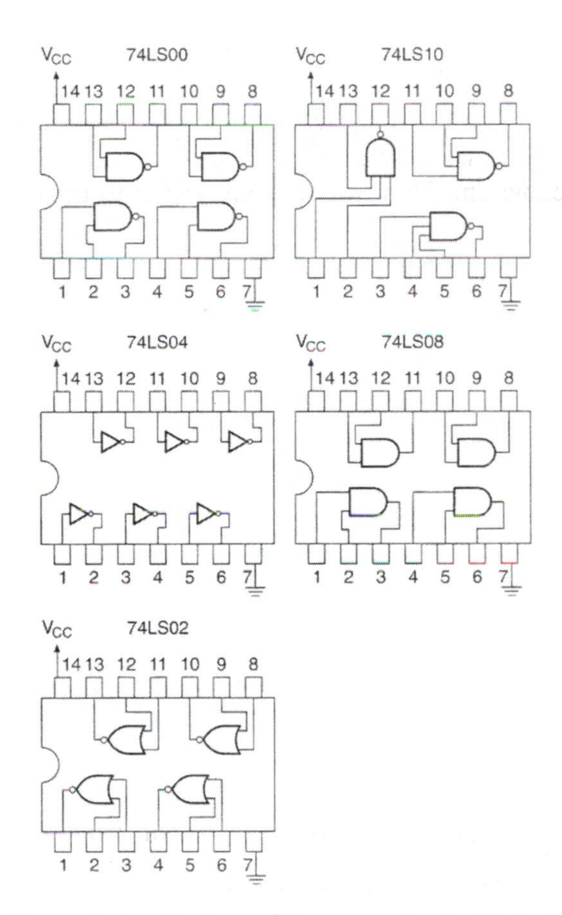

Figure 3.4 Pinouts of Common Logic Gate ICs

Assignment Questions

1. In the experimental notes for this laboratory exercise, it was stated that the circuits in Figures 3.1 and 3.2 are equally efficient to build. That is, each circuit requires the same number of logic gate packages to build. Prove that this is true by listing the IC packages, with part numbers, required to build each circuit.
2. Complete problems 3.6, 3.7, 3.22, and 3.24 in the textbook.

Standard Wiring for Altera UP-1 and RSR PLDT-2 Boards

The CPLD labs in this manual have been designed for use with either an Altera UP-1 CPLD circuit or an RSR PLDT-2 board. Other CPLD boards with an Altera EPM7128SLC84 CPLD can also be used, but may or may not conform to the standard wiring configuration designed for the labs. In this case, you will need to assign pins to your CPLD designs according to the layout of your board.

Labs 4 through 15 in this manual are designed with a standard wiring configuration, so that the board can be wired once and reprogrammed as necessary for the practical exercises in the various labs. The configuration is shown in the table on the page 18. My recommendation is to spend half an hour or so making the connections as directed, then leave the board alone until you are ready for Labs 16 and 17, at which time some minor changes will need to be made.

For the Altera UP-1 board, the connections should be as indicated on page 18. For the RSR PLDT-2 board, connections to SW1-1 through SW1-8 and to LED1 through LED8 are normally made by removable shorting jumpers. If the jumpers are left in place, further wiring is not required for these inputs and outputs.

The standard board connections can be tested by downloading a test routine in a file called **board_test_2.gdf** (Altera UP-1) or **board_test_3.gdf** (RSR PLDT-2). These files are available as uncompiled design files and components to students and as compiled files to instructors.

To use a student file, copy the folder **\student files\board** to your working drive and store it as a subfolder of **\maxzwork**. Open MAX+PLUS II, then either **board_test_2.gdf** or **board_test_3.gdf.** From the **File, Project** menu, choose **Set Project to Current File,** then **Save and Compile.** When the compile process is complete, choose **Programmer** from the **MAX+PLUS II** menu. Click **Program.** (This assumes that the CPLD board is connected to the PC parallel port via a ByteBlaster or equivalent cable.)

The test files cause an increasing hexadecimal count to appear on the board numerical displays. The files also test the DIP switch and LED connections by turning on an LED when a corresponding switch is made HIGH. The switches and LEDs operate in the same order as they are laid out on the board (i.e., the first switch controls the first LED).

EPM7128LC84-7 Pin Assignments
Altera UP-1 Board and PLDT-2 Board

Seven Segment Digits			
Function	**Pin**	**Function**	**Pin**
a1	58	a2	69
b1	60	b2	70
c1	61	c2	73
d1	63	d2	74
e1	64	e2	76
f1	65	f2	75
g1	67	g2	77
dp1	68	dp2	79

Pushbuttons			
Function	**Pin**	**Function**	**Pin**
PB1	11	PB2	1

DIP Switches			
Function	**Pin**	**Function**	**Pin**
SW1-1	34	SW2-1	28
SW1-2	33	SW2-2	29
SW1-3	36	SW2-3	30
SW1-4	35	SW2-4	31
SW1-5	37	SW2-5	57
SW1-6	40	SW2-6	55
SW1-7	39	SW2-7	56
SW1-8	41	SW2-8	54

LED Outputs			
Function	**Pin**	**Function**	**Pin**
LED1	44	LED9	80
LED2	45	LED10	81
LED3	46	LED11	4
LED4	48	LED12	5
LED5	49	LED13	6
LED6	50	LED14	8
LED7	51	LED15	9
LED8	52	LED16	10

Unassigned: Pins 12, 15, 16, 17, 18, 20, 21, 22, 24, 25, 27

Special Function: Pin 1 (GCLRn); Pin 2 (Input/OE2/GCLK2);
Pin 83 (GCLK1, hardwired); Pin 84 (OE1)

Shaded cells in table do not need to be wired for the RSR PLDT-2 board if shorting jumpers are installed. Seven-segment displays do not need to be wired on either the Altera UP-1 or the RSR PLDT-2 board.

Lab 4

Introduction to MAX+PLUS II

Name ___________________________________ Class ___________ Date ________________

Objectives Upon completion of this laboratory exercise, you should be able to:

- Enter a simple logic circuit using the MAX+PLUS II Graphic Editor.
- Compile a MAX+PLUS II design file.
- Download the file to an Altera CPLD on the Altera UP-1 board or RSR PLDT-2 board.

Reference Dueck, Robert K., *Digital Design with CPLD Applications and VHDL*

Chapter 4: Introduction to PLDs and MAX+PLUS II
4.1 What is a PLD?
4.2 Programming PLDs using MAX+PLUS II
4.3 Graphic Design File
4.4 Compiling MAX+PLUS II Files
4.7 Creating a Physical Design

Equipment Required

CPLD Trainer:
Altera UP-1 Circuit Board with ByteBlaster Download Cable, or
RSR PLDT-2 Circuit Board with Straight-Through Parallel Port Cable, or
Equivalent CPLD Trainer Board with Altera EPM7128S CPLD
MAX+PLUS II Student Edition Software
AC Adapter, minimum output: 7 VDC, 250 mA DC
Anti-static wrist strap
#22 solid-core wire
Wire strippers

Experimental Notes

Programmable logic devices (PLDs) are digital integrated circuits that do not have permanently defined functions, but are programmed by the end user. This is in contrast to the logic ICs used in previous labs that have fixed logic functions.

A Complex PLD (CPLD) can be programmed using special software such MAX+PLUS II by Altera Corporation. Digital circuits can be entered in MAX+PLUS II as a schematic or using a text-based Hardware Description Language (HDL) and then programmed into a CPLD through a connection to the parallel port of the PC running MAX+PLUS II.

In this lab we will use the schematic entry capabilities of MAX+PLUS II to enter a simple combinational logic circuit and program it into a PLD. The circuit we will use is the majority vote circuit, shown in Figure 4.1. This circuit generates a HIGH output when at least two out of three inputs are HIGH.

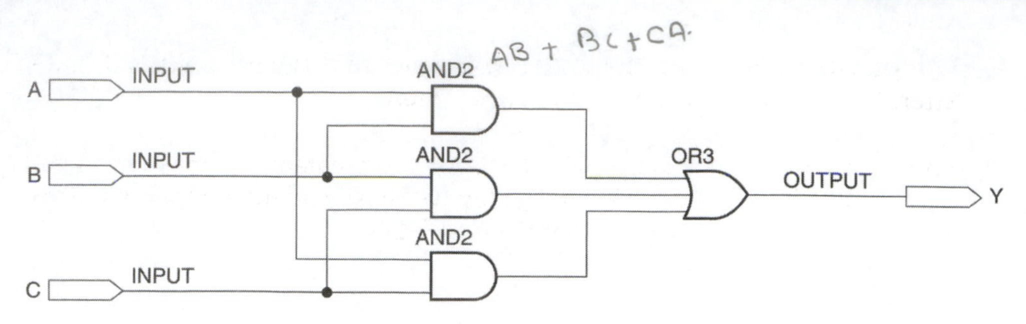

Figure 4.1 MAX+PLUS II Graphic Design File of a Majority Vote Circuit

MAX+PLUS II automatically generates a number of files to keep track of the PLD programming information represented by the Graphic Design File. These files, taken together, represent a **project** in MAX+PLUS II. All operations required to create a programming file for a CPLD are performed on a project, not a file. Thus, it is important, during the design process, to keep track of what the current project is. The MAX+PLUS II toolbar, shown in Figure 4.2, makes this fairly easy.

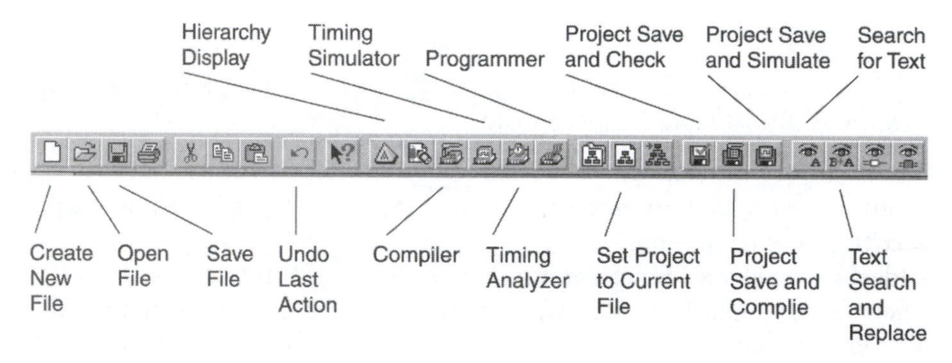

Figure 4.2 MAX+PLUS II Toolbar

The toolbar has a number of buttons that pertain to the current project of a PLD design. The operations performed by these buttons can all be done through the regular menus of MAX+PLUS II, but the toolbar offers a quick way to access many available functions. You can find out the function of any button by placing the cursor on the button and reading a description at the bottom of the window.

In particular, notice the buttons that create, open, and save files (standard Windows icons) and the button that sets the project to the current file. When creating a new file, make it standard practice to first **Save** the file, then **Set Project to Current File.** If you do this as a habit, you (and MAX+PLUS II) will always know what the current project is. If you don't, you will find that you are saving or compiling some other project and wondering why your last set of changes didn't work. You can also set the project to the current file through the **Project** submenu of the **File Menu.**

Another good practice is to create a new Windows folder for each new design that you enter. Since MAX+PLUS II creates many files in the design process, the folders would become unmanageable if designs were not kept in separate folders.

MAX+PLUS II installs a folder for working with design files called **max2work.** The examples in this text will be created in a subfolder of **max2work.** If you are working in a situation where many people share a computer and you have access to a network drive of your own, you may wish to keep your working files in a **max2work** folder on the network drive. Avoid storing your working files on a local hard drive unless you are the only one with regular access to the computer. Examples in this lab manual will not specify a drive letter, but will indicate ***drive:*\max2work*folder***.

Note Although the examples in this book are created with the Altera UP-1 board or the RSR PLDT-2 board in mind, they will easily adapt to other circuit boards carrying an Altera EPM7128S or other similar CPLD.

Procedure

Write the Boolean equation of the majority vote circuit shown in Figure 4.1.

$Y =$ AB+BC+CA.

Start MAX+PLUS II and enter the schematic of the majority vote circuit by following the procedure listed below.

Entering Components

To create a Graphic Design File, click the **New File** icon on the tool bar or choose **New** on the MAX+PLUS II **File** menu. The dialog box, shown in Figure 4.3 appears. Select **Graphic Editor file** and choose OK.

Maximize the window and click the **Save** icon or choose **Save As** or **Save** from the **File** menu. In the dialog box shown in Figure 4.4, save the file in a new folder (eg., ***drive:\ max2work\maj_vote\maj_vote.gdf***) and choose **OK.** (If you have not created the new folder, just type the complete path name in the **File Name** box. MAX+PLUS II will create a new folder.) Click the icon to **Set Project to Current file** or choose this action from the **File, Project** menu.

The first design step is to lay out and align the required components. We require three 2-input AND gates, a 3-input OR gate, three input pins, and one output pin. These basic components are referred to as **primitives.** Let us start by entering three copies of the AND gate primitive, called **and2.**

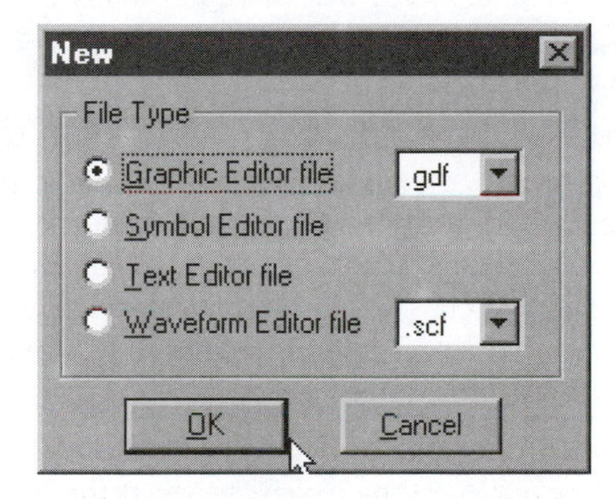

Figure 4.3 New Dialog Box

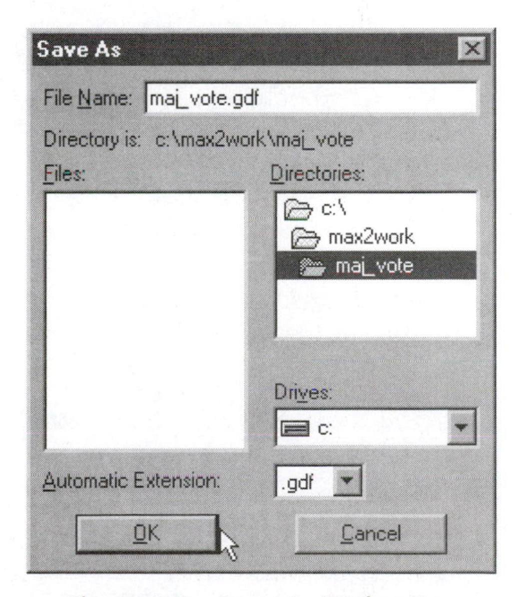

Figure 4.4 Save As Dialog Box

Click the left mouse button to place the cursor (a flashing square) somewhere in the middle of the active window. Right-click to get a pop-up memu, shown in Figure 4.5, and choose **Enter Symbol**. The dialog box in Figure 4.6 appears. Type **and2** in the **Symbol Name** box and choose **OK**. A copy or **instance** of the and2 primitive appears in the active window.

You can repeat the above procedure to get two more instances of the and2 primitive, or you can use the **Copy** and **Paste** commands. These are the same icons and **File** commands as for other Windows programs. Highlight the **and2** symbol by clicking it. Right-click the symbol to get the pop-up menu shown in Figure 4.7 and choose **Copy**.

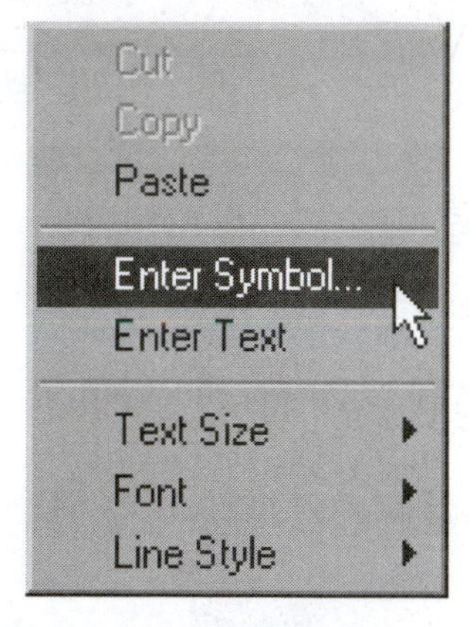

Figure 4.5 Enter Symbol Pop-up Menu

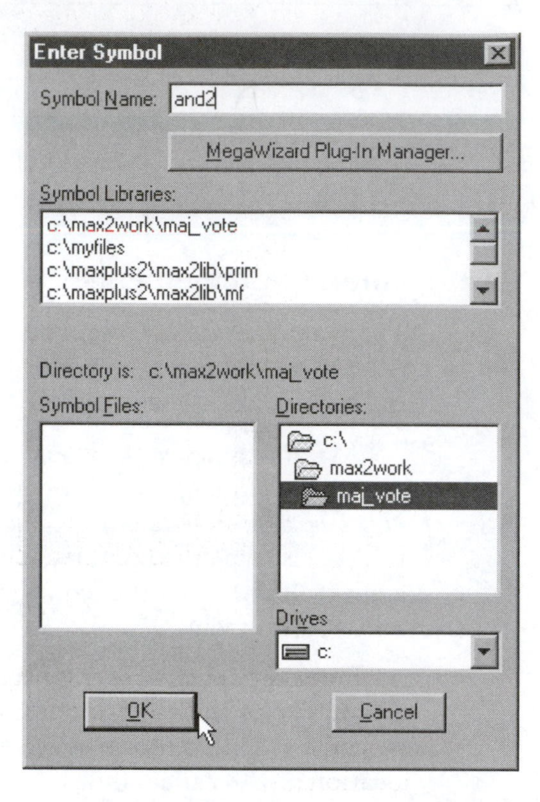

Figure 4.6 Enter Symbol Dialog Box

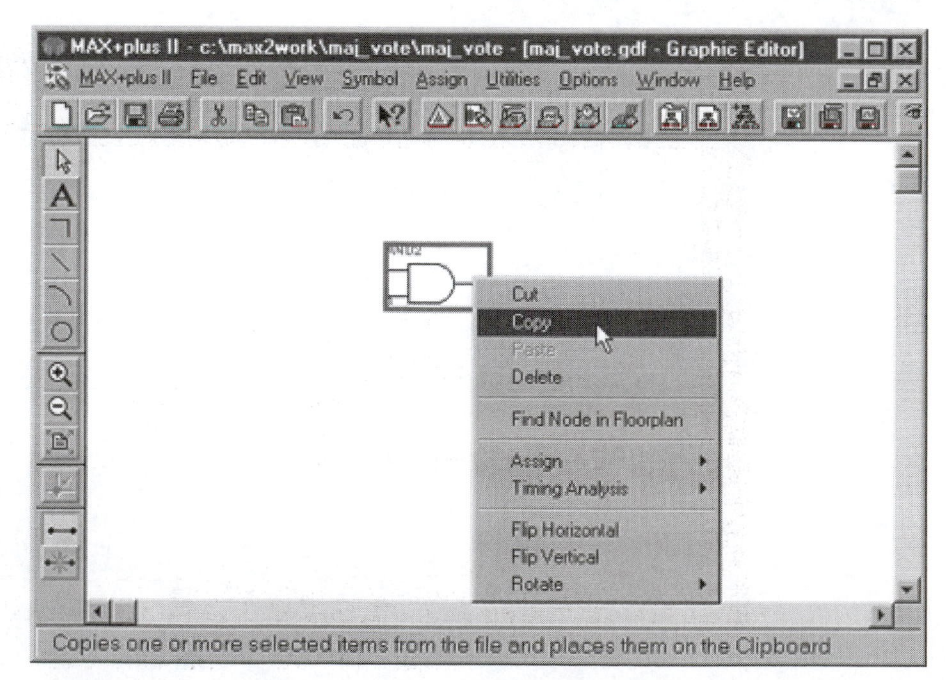

Figure 4.7 Copying a Component

You can also click the **Copy** icon on the toolbar or use the **Copy** command in the **File** menu.

Paste an instance of the primitive by clicking to place the cursor, then right-clicking to sbring up the menu shown in Figure 4.8. Choose **Paste**. The component will appear at the cursor location, marked in Figure 4.8 by the square at the top left corner of the pop-up menu.

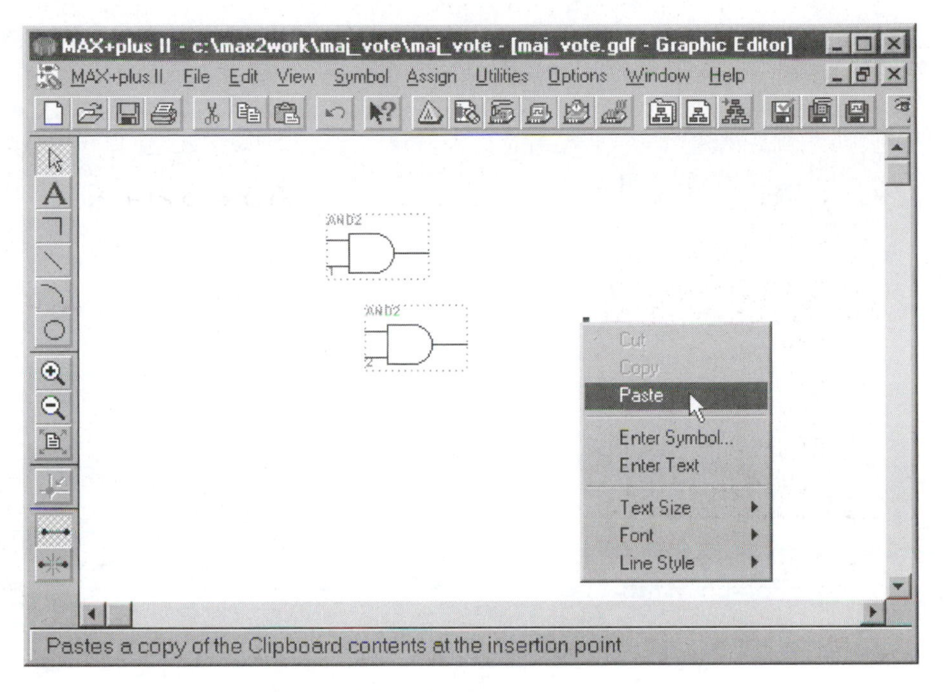

Figure 4.8 Pasting a Component

Enter the remaining components by following the **Enter Symbol** procedure outlined above. The primitives are called **or3**, **input**, and **output**. When all components are entered we can align them, as in Figure 4.9 by highlighting, then dragging each one to a desired location.

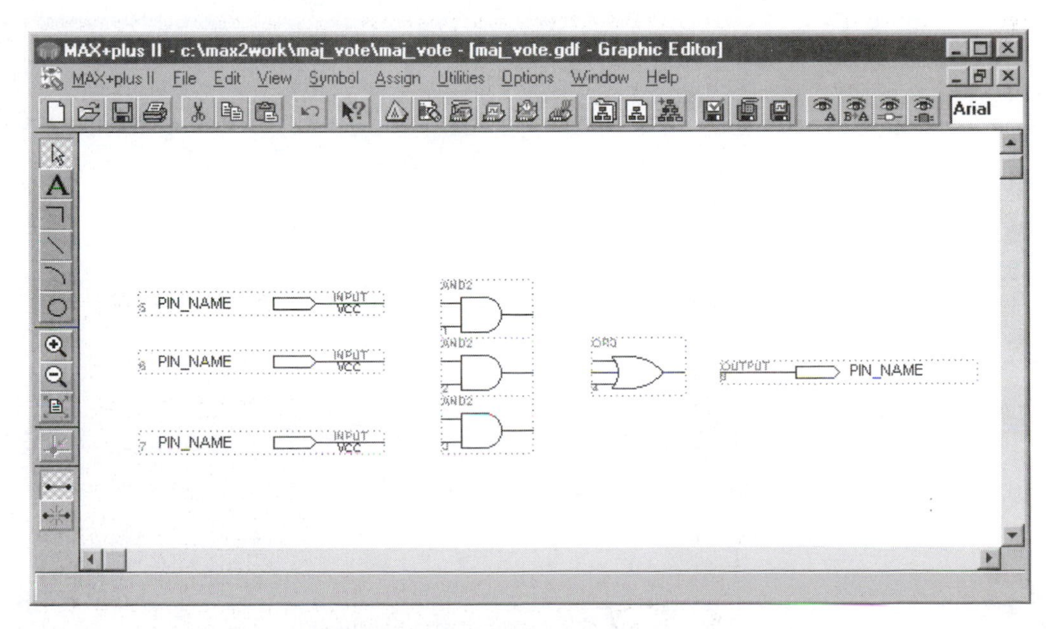

Figure 4.9 Aligned Components

Connecting Components

To connect components, click over one end of one component and drag a line to one end of a second component. When you drag the line, a horizontal and vertical broken line mark the cursor position, as shown in Figure 4.10. These lines help you align connections properly.

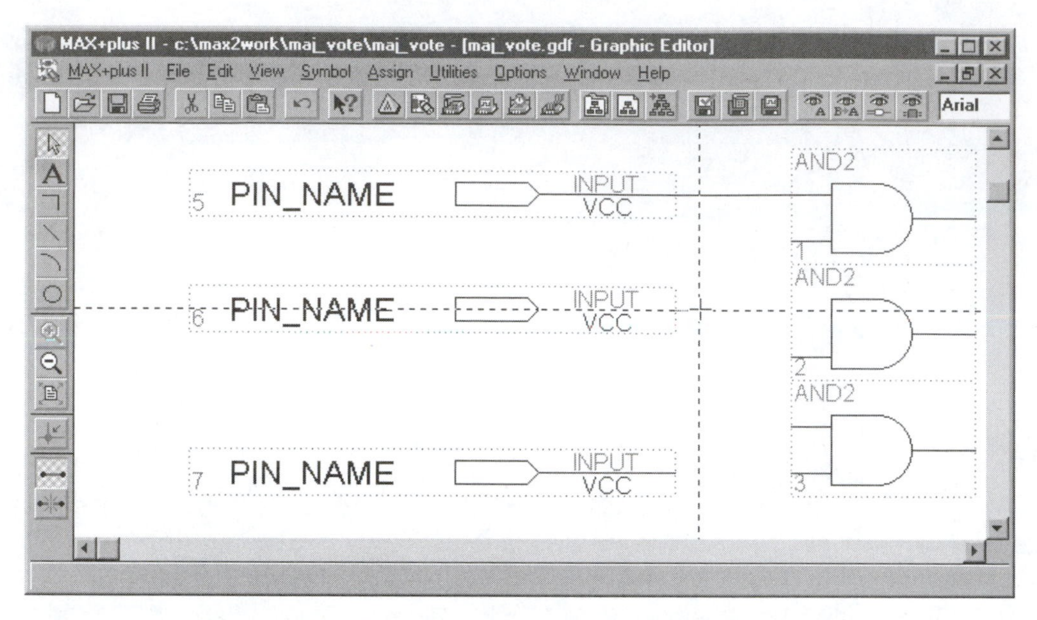

Figure 4.10 Dragging a Line to Connect Components

A line will automatically make a connection to a perpendicular line, as shown in Figure 4.11.

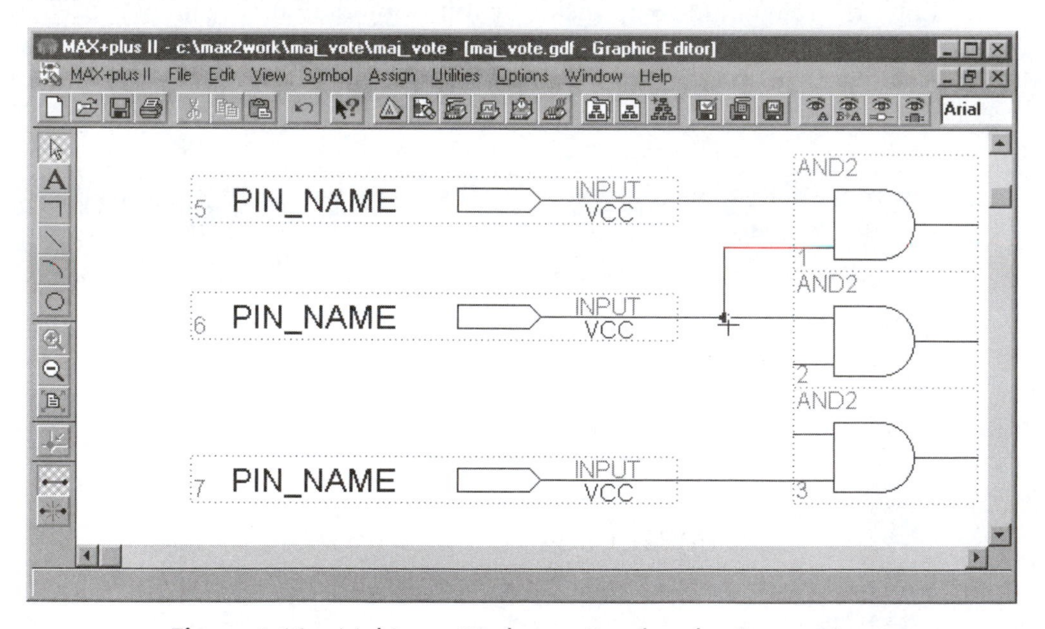

Figure 4.11 Making a 90-degree Bend and a Connection

A line can have one 90-degree bend, as at the inputs of the AND gates. If a line requires two bends, such as shown at the AND outputs in Figure 4.12, you must draw two separate lines.

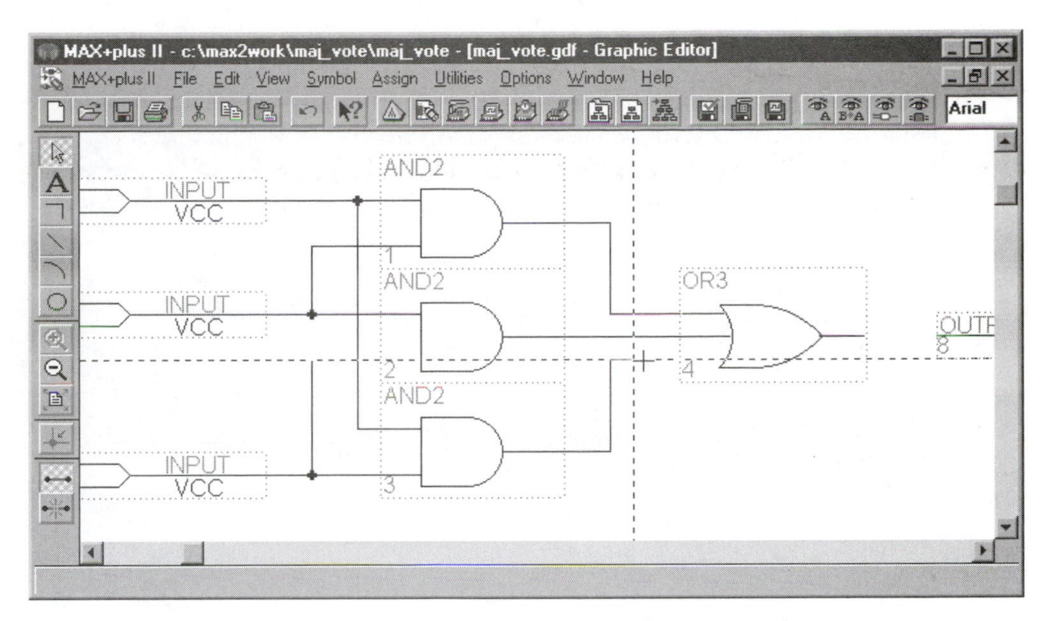

Figure 4.12 Line with Two 90-degree Bends

Assigning Pin Names

Before a design can be compiled, its inputs and outputs must be assigned names. We could also specify pin numbers, if we wished to make the design conform to a particular CPLD, but it is not necessary to do so at this stage. It may not even be desirable to assign pin numbers, since the design we enter can be used as a component or subdesign of a larger circuit. We may also wish MAX+PLUS II to assign pins to make the best use of the CPLD's internal resources. At any rate, we will leave this step out for now.

Figure 4.13 shows the naming procedure. Pins A and B have already been assigned names. Highlight a pin by clicking on it. Right-click the highlighted pin and choose **Edit Pin Name** from the pop-up menu. You could also double-click the pin name to highlight it. Type in the new name.

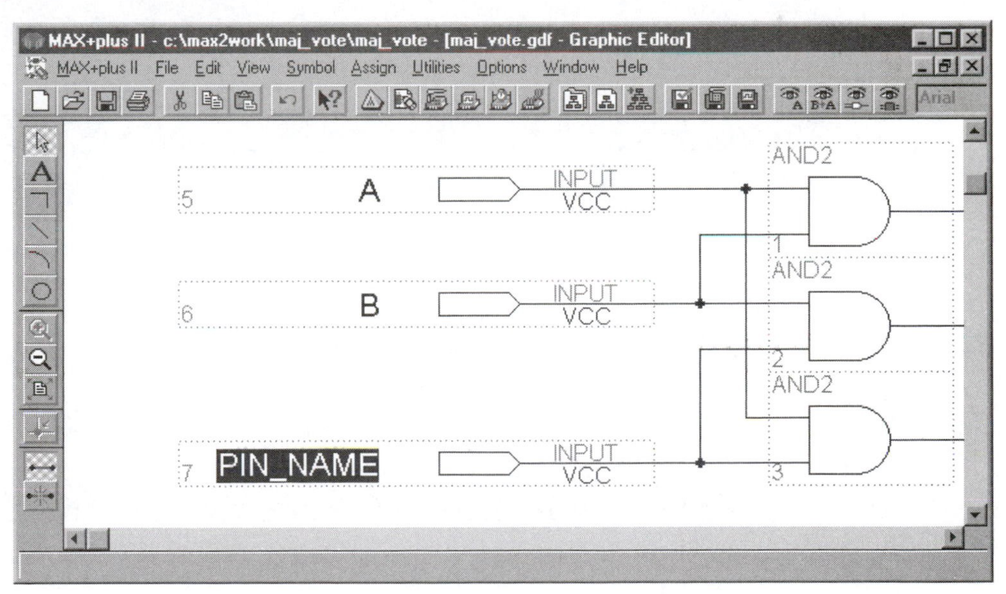

Figure 4.13 Assigning Pin Names

If there are several pins that are spaced one above the other, you can highlight the top pin name, as described above, the highlight successive pin names by using the **Enter** key.

Compiling MAX+PLUS II Files

The MAX+PLUS II compiler converts design entry information into binary files that can be used to program a PLD. Before compiling, we should assign a target device to the design.

From the **Assign** menu, shown in Figure 4.14, select **Device.** From the dialog box in Figure 4.15, select the target device. For the Altera UP-1 board, this would be either the EPM7128SLC84-7 (shown) or the FLEX10K20RC240-4. For the RSR PLDT-2 board, choose EPM7128SLC84-15. The device family for the EPM7128S device is MAX70000S.

Note To see the EPM7128SLC84-7 or EPM7128SLC84-15 device, the box that says **Show Only Fastest Speed Grades** must be unchecked.

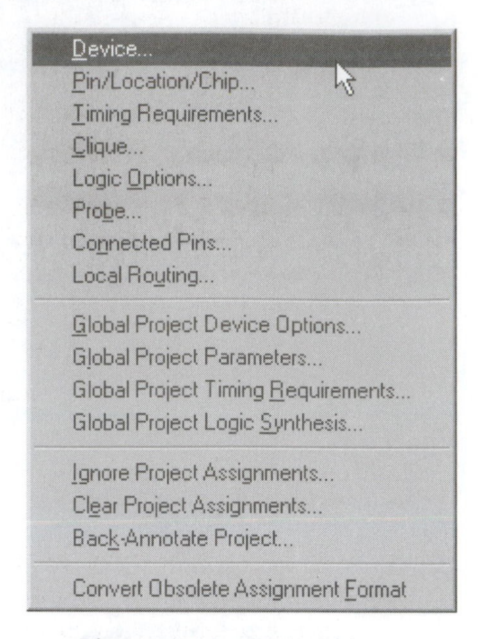

Figure 4.14 Assign Menu

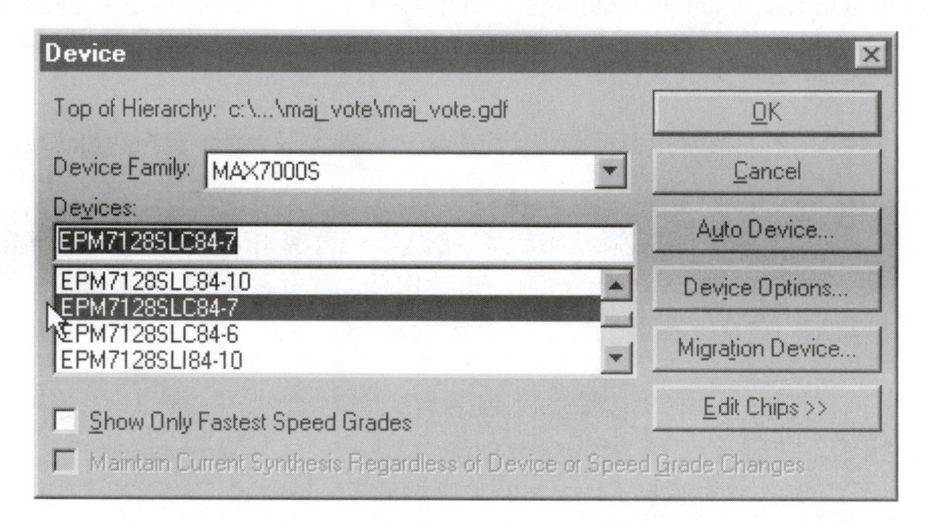

Figure 4.15 Device Dialog Box

The compiler has a number of settings that can be chosen prior to the actual compile process. Figure 4.16 shows some of the settings that should be selected from the **Processing** menu of the **Compiler** window. You can open the **Compiler** window from the **MAX+PLUS II** menu or by clicking the **Compiler** button on the toolbar at the top of the screen.

Design Doctor is a utility that checks for adherence to good design practice and will warn you of any bad design choices. (Design Doctor will not stop the design from compiling, but will suggest potential problems that could result from a particular design.) The **Timing SNF Extractor** creates a Simulation Netlist File, which is required to perform a timing simulation of the design. We will perform this step in later MAX+PLUS II designs. (If you are not able to select the **Timing SNF Extractor**, then uncheck the **Functional SNF Extrator** option.) **Smart Recompile** allows the compiler to use previously compiled portions of the design to which no changes have been made. This allows the

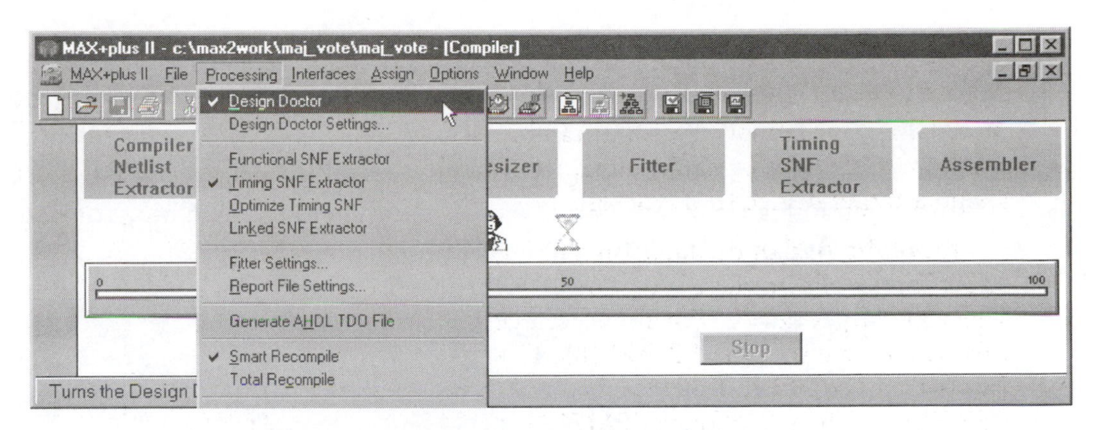

Figure 4.16 MAX+PLUS II Compiler Settings

compiler to avoid having to compile the entire design each time a change is made to one part of the design.

To start the compile process, click **Start** in the **Compiler** window. While in progress, the window will look something like Figure 4.17. Message of three types may appear during the compile process. **Info** messages (green text) are for information only. **Warning** messages (blue text) tell you of potential, but nonfatal, problems with the design. **Error** messages (red text) inform you of design flaws that render the design unusable. A PLD can still be programmed if the compiler generates **info** or **warning** messages, but not if it generates an **error**.

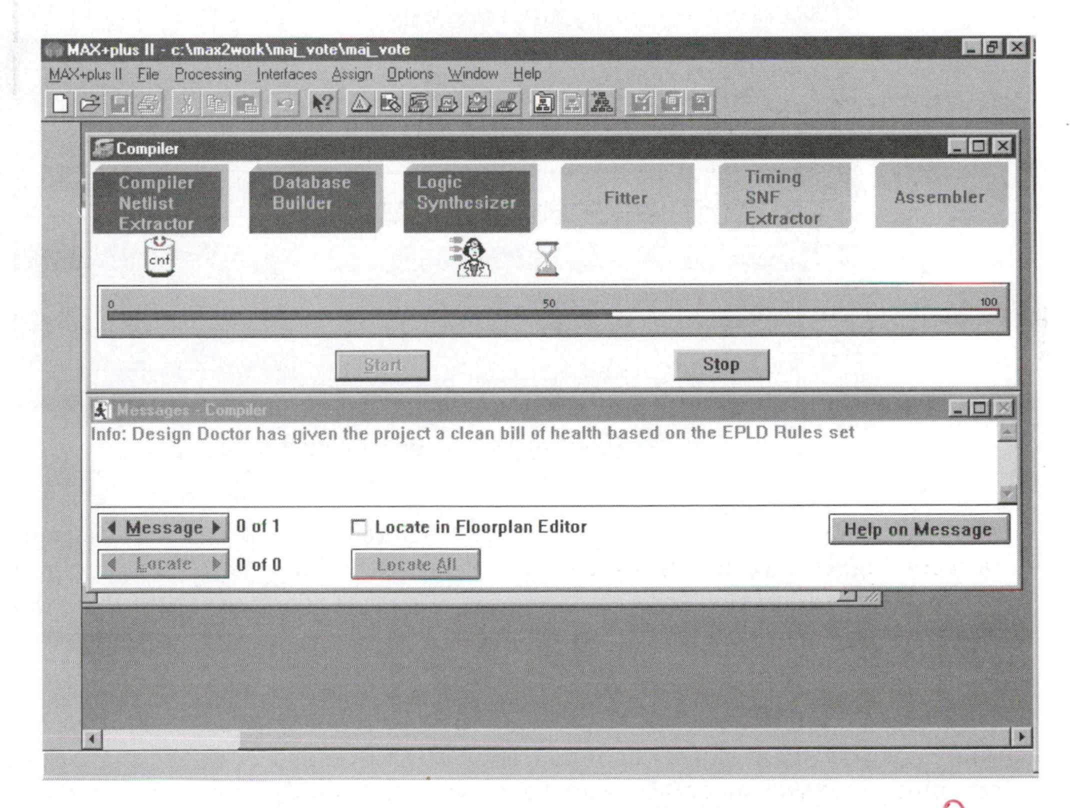

Figure 4.17 MAX+PLUS II Compiler Operation

Compile the design. Show your instructor before proceeding.

Instructor's Initials: ________

Assigning Pin Numbers

Before proceeding with this step, make sure that you have assigned a device part number to the design. Save the file and set the project to the current file.

To assign a pin number, click on the pin to highlight it, then right-click to see the pop-up menu in Figure 4.18. Choose **Assign**, then **Pin/Location/Chip.** You can also do this from the **Assign** menu at the top of the screen.

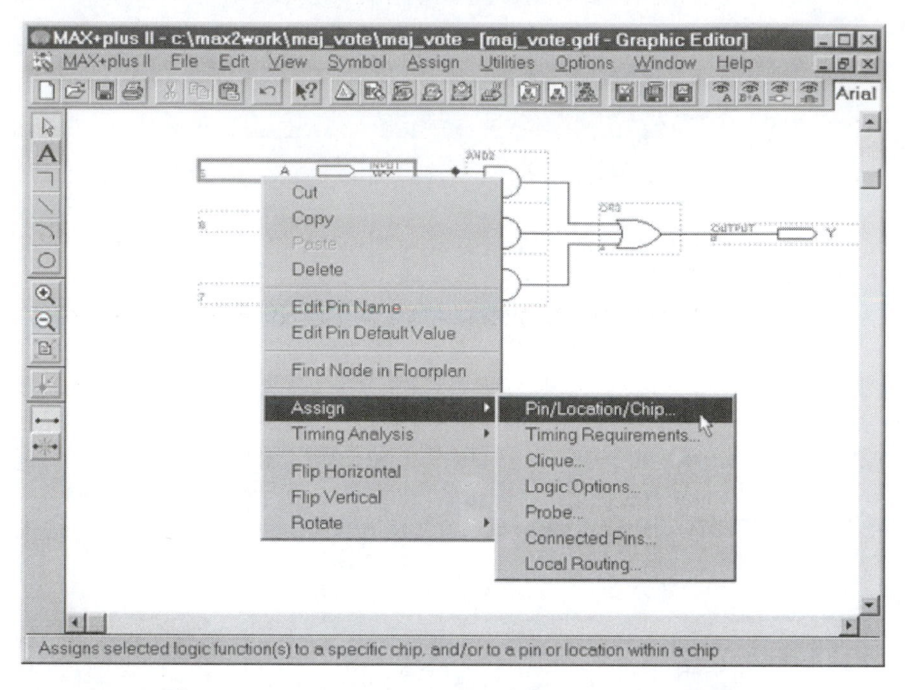

Figure 4.18 Pop-up Menu for Pin Assignments

We can assign pin numbers in the dialog box in Figure 4.19.

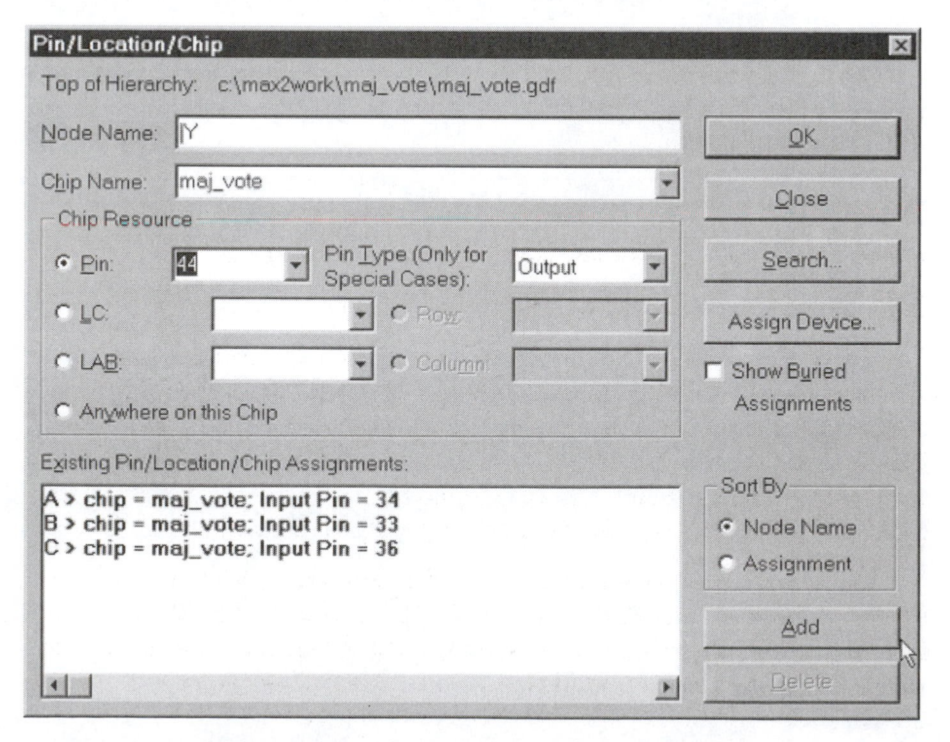

Figure 4.19 Pin/Location/Chip Dialog Box

Type **A** in the **Node Name** box, **34** in the **Pin** box and click **Add.** Type **B** in the **Node Name** box, assign this name to pin 33, and click **Add.** Repeat this procedure until all names are assigned, as in Table 4.1. When all assignments are complete, click **OK.**

Table 4.1 Pin Assignments for Majority Vote Circuit

Pin Name	Pin Number
A	34
B	33
C	36
Y	44

Figure 4.20 shows the input pin assignments as they appear in the **gdf** file.

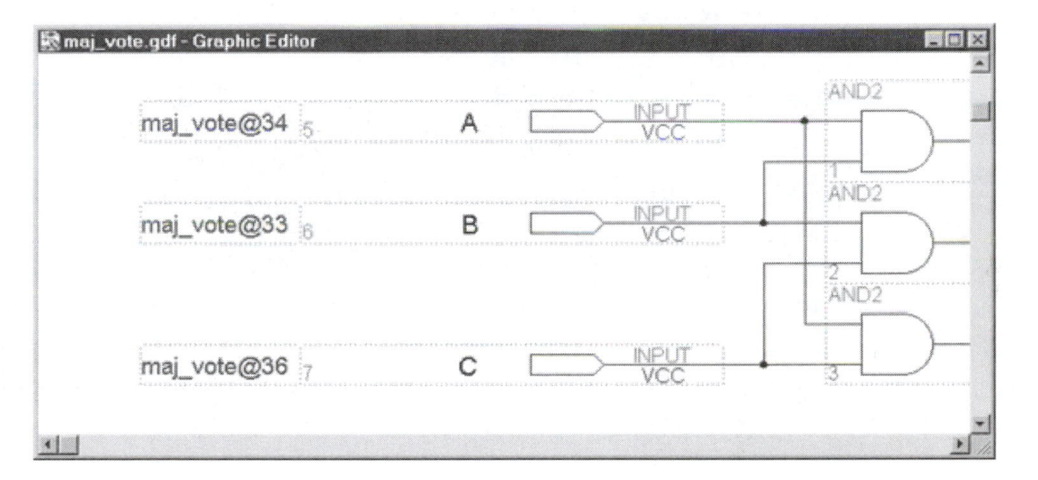

Figure 4.20 Pin Assignments As Seen in a gdf File

Set Project to Current File (File menu) and compile again. Show your instructor before proceeding.

Instructor's Initials: ________

Programming CPLDs on the Altera UP-1 Circuit Board

(Skip this section if you are using an RSR PLDT-2 or other circuit board.)

The CPLDs on the Altera UP-1 circuit board are programmed via the programming software in MAX+PLUS II and a ribbon cable called the **ByteBlaster**. The ByteBlaster, shown in Figure 4.21, connects the parallel port of a the PC running MAX+PLUS II to a 10-pin male socket that complies with the **JTAG** standard. This standard specifies a four-wire interface, originally developed for testing chips without removing them from a circuit board, but can also be used to program or configure PLDs.

To program a device on the Altera UP-1 board, set the jumpers to program the EPM7128S or configure the EPF10K20, as shown in the *Altera University Program Design Laboratory Package User Guide.* Connect the ByteBlaster cable from the parallel port of the PC running MAX+PLUS II to the 10-pin JTAG header. (You may have to run a 25-wire cable (male-D-connector-to-female-D-connector) to make it reach.) Plug an AC adapter (9-volt dc output) into the power jack of the UP-1 board.

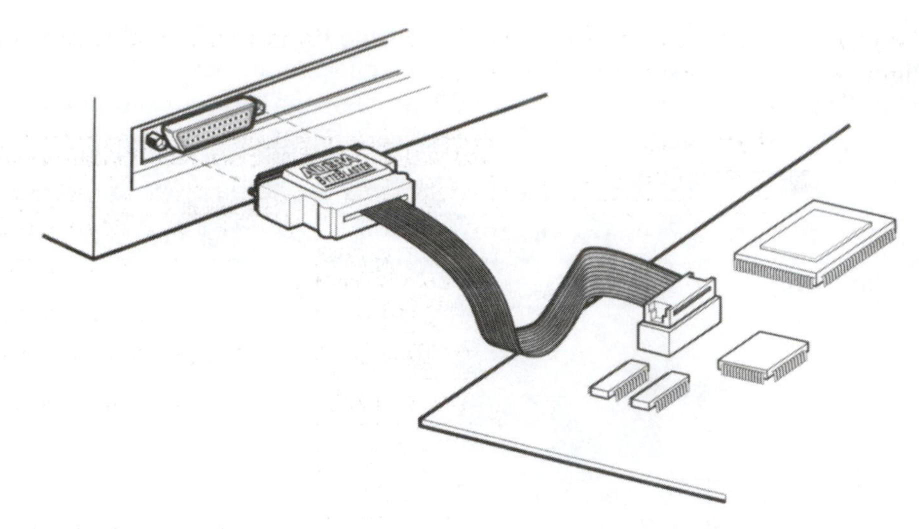

Figure 4.21 ByteBlaster Parallel Port Download Cable

Programming CPLDs on the RSR PLDT-2 Circuit Board

(Skip this section if you are using an Altera UP-1 board.)

The CPLDs on the **RSR PLDT-2** circuit board are programmed via the programming software in MAX+PLUS II and a cable that connects to the parallel port of a the PC running MAX+PLUS II. The board contains circuitry that implements a JTAG port, a four-wire interface used for testing and programming CPLDs and other JTAG-compliant devices.

Plug an AC adapter (9-volt dc output) into the power jack of the UP-1 board and the supplied parallel port cable to the parallel port of your PC.

Programming the CPLD

Open the file you wish to download to the UP-1 board (e.g., **maj_vote.gdf**). Set the project to the current file. Invoke the MAX+PLUS II Programmer from the **MAX+PLUS II** menu or click the Programmer button (the icon showing the blue ribbon cable) on the MAX+PLUS II toolbar.

If you have never programmed a device with your copy of MAX+PLUS II, you will need to set up the hardware configuration. Click **Hardware Setup** in the **Options** menu to get the dialog box in Figure 4.22.

Select **ByteBlaster** in the **Hardware Type** box. Ensure that **Parallel Port** is the same as the port the ByteBlaster is plugged into (usually LPT1:). Click **OK**. (If you have a choice, configure your parallel port as an Enhanced Communications Port (ECP) in your computer's CMOS setup. For most users this step is not necessary, as the port is already configured this way.)

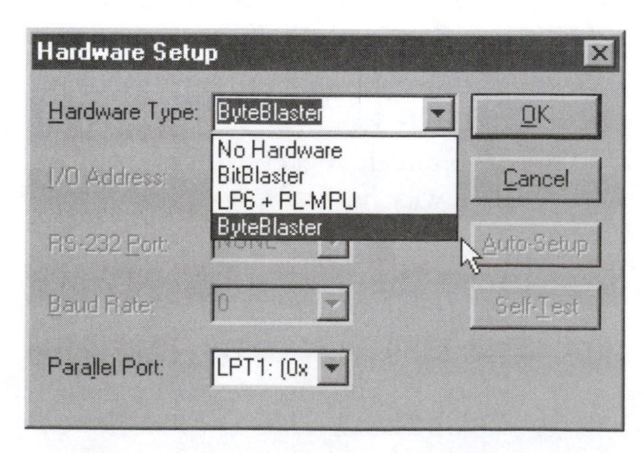

Figure 4.22 Hardware Setup Dialog Box

If the current project was compiled with the MAX7000S device selected, the **pof** (Programmer Object File) for the project will automatically be available. The **Programmer** dialog box will appear as in Figure 4.23. To download, click **Program.**

Figure 4.23
Programmer Dialog Box (MAX7000S Device)

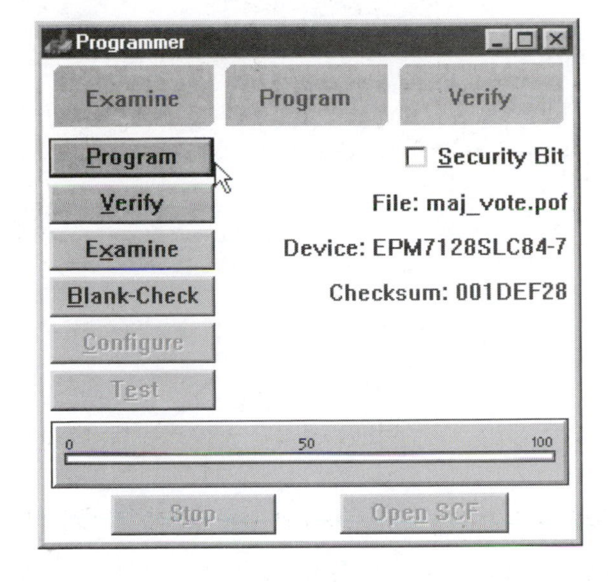

Simulation done.
03/18/05

Instructor's Initials: ______

Wiring the Altera UP-1 Board

(Skip this section if you are using an RSR PLDT-2 circuit board.)

Connect short lengths of #22 wire from three DIP switches on the UP-1 board to pins 34, 33, and 36 on the MAX prototyping headers on the Altera UP-1 board. Connect a wire from pin 44 to an LED connector on the UP-1 board. Take the truth table of the programmed design by making all combinations of inputs A, B, and C and noting the status of the output LED for each combination. (DIP switches: UP = 1, DOWN = 0; LED: ON = 0, OFF = 1.)

Instructor's Initials: ______
03/18/05

Truth Table:

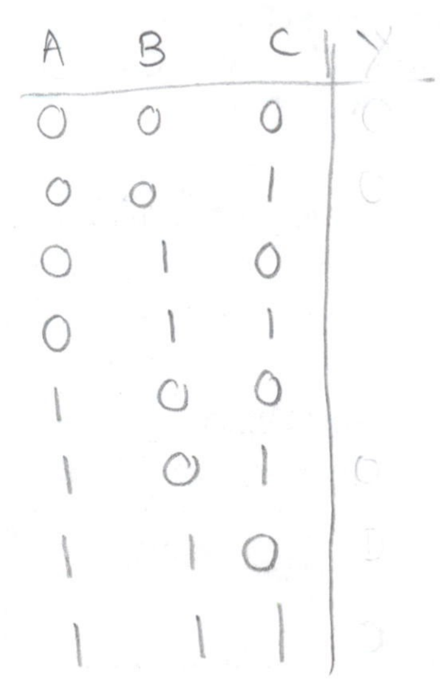

A	B	C
0	0	0
0	0	1
0	1	0
0	1	1
1	0	0
1	0	1
1	1	0
1	1	1

Wiring the RSR PLDT-2 Board

(Skip this section if you are using an Altera UP-1 circuit board.)

No wiring is necessary for the RSR PLDT-2 board, as the required connections from the DIP switches and the LED to the CPLD are made via jumpers installed on the board.

Take the truth table of the programmed design by making all combinations of inputs A, B, and C and noting the status of the output LED for each combination. (LED: ON = 1, OFF = 0.)

Instructor's Initials: __________

Truth Table:

Note A major difference between the Altera UP-1 board and the RSR PLDT-2 board is that the LEDs and numerical displays on the **Altera UP-1 board** are **active-LOW** (illuminated by logic 0) and the LEDs and numerical displays on the **RSR PLDT-2** board are **active-HIGH** (illuminated by logic 1). If you are using the Altera UP-1 board, you must remember that an ON LED represents a 0 or you will have to invert all circuit outputs to make ON LEDs indicate logic HIGH levels.

Assignment Questions

1. Write the unsimplified Boolean expression of the majority vote circuit in sum-of-products form. Use Boolean algebra to simplify the expression as much as possible.
2. Complete problems 4.1 through 4.7 in Chapter 4 of the textbook.

Lab 5

Introduction to VHDL

Name ______________________ Class ________ Date __________

Objectives Upon completion of this laboratory exercise, you should be able to:

- Enter a simple combinational logic circuit in VHDL using the MAX+PLUS II Text Editor.
- Assign a target device and pin numbers and compile a VHDL design file.
- Download the file to an Altera CPLD on the Altera UP-1 board or RSR PLDT-2 board.

Reference Dueck, Robert K., *Digital Design with CPLD Applications and VHDL*

Chapter 4: Introduction to PLDs and MAX+PLUS II;
4.6 Text Design File (VHDL)

Equipment Required

CPLD Trainer:
Altera UP-1 Circuit Board with ByteBlaster Download Cable, or
RSR PLDT-2 Circuit Board with Straight-Through Parallel Port Cable, or
Equivalent CPLD Trainer Board with Altera EPM7128S CPLD
MAX+PLUS II Student Edition Software
AC Adapter, minimum output: 7 VDC, 250 mA DC
Anti-static wrist strap
#22 solid-core wire
Wire strippers

Experimental Notes

VHDL stands for VHSIC Hardware Description Language. (VHSIC = Very High Speed Integrated Circuit.) VHDL is an industry standard programming language for simulation and synthesis of digital circuits. In this lab, we will use VHDL to enter the designs for some simple combinational logic circuits, using VHDL constructs for Boolean equations and truth tables.

Every VHDL design requires an **entity declaration**, which describes the inputs and outputs of the design, and an **architecture body**, which describes the internal relationship between inputs and outputs. Within the architecture body, we can use many different language statements to describe our design. We will use two of the simplest: a **concurrent signal assignment statement**, which can be used to implement a Boolean expression, and a **selected signal assignment statement**, which can be used, among other things, to implement a truth table.

Figure 5.1 shows the majority vote circuit constructed in Lab 4. The circuit has the Boolean expression $Y = A\,B + B\,C + A\,C$. The output is HIGH if at least two out of three

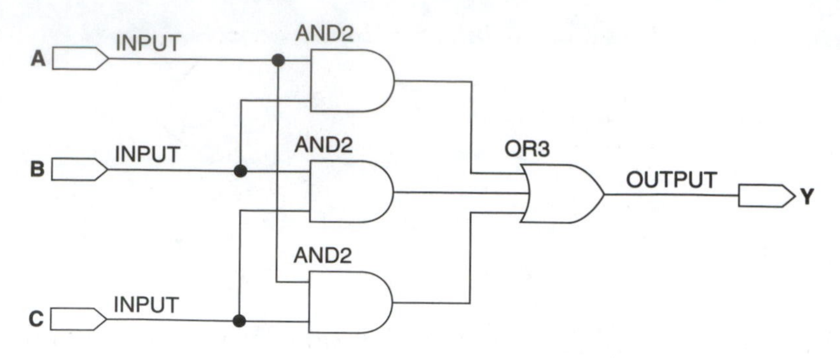

Figure 5.1 Majority Vote Circuit

inputs are HIGH. A concurrent signal assignment statement would encode this equation as follows:

```
y <= (a and b) or (b and c) or (a and c);
```

The complete VHDL file is as follows:

```
-- maj_vot2.vhd
-- Majority vote circuit using a concurrent signal assignment statement
LIBRARY ieee;
USE ieee.std_logic_1164.ALL;

ENTITY maj_vot2 IS
PORT(
    a, b, c    : IN     STD_LOGIC;
    y          : OUT    STD_LOGIC);
END maj_vot2;

ARCHITECTURE a OF maj_vot2 IS
BEGIN
    y <= (a and b) or (b and c) or (a and c);
END a;
```

Refer to pages 135–137 in *Digital Design with CPLD Applications and VHDL* for a more detailed description of the various parts of this design file.

Encoding a Truth Table in VHDL

We can encode a truth table in VHDL by using a selected signal assignment statement. Examine the truth table in Table 5.1.

In the entity declaration of a VHDL file, we can define the combined input values of D_1 and D_0 as a 2-bit vector of type STD_LOGIC_VECTOR:

```
d: IN STD_LOGIC_VECTOR (1 downto 0);
```

Table 5.1 Sample Truth Table

D_1	D_0	Y
0	0	0
0	1	1
1	0	1
1	1	0

This truth table can be encoded by the following selected signal assignment statement:

```
WITH d SELECT
   y    <=  '0' WHEN "00",
            '1' WHEN "01",
            '1' WHEN "10",
            '0' WHEN "11",
            '0' WHEN others;
```

The output value is shown on the left side of the each line of the statement and the corresponding input combinations on the right side. We include the **others** clause because the STD_LOGIC_VECTOR inputs are not fully specified by combinations of '0' and '1'. The **others** clause specifies the default value of the output.

The complete VHDL file is as follows:

```
-- xor2.vhd
-- XOR truth table using a selected signal assignment statement
LIBRARY ieee;
USE ieee.std_logic_1164.ALL;

ENTITY xor2 IS
PORT(
   d: IN     STD_LOGIC_VECTOR (1 downto 0);
   y: OUT    STD_LOGIC);
END xor2;

ARCHITECTURE a OF xor2 IS
BEGIN
WITH d SELECT
   y    <=  '0' WHEN "00",
            '1' WHEN "01",
            '1' WHEN "10",
            '0' WHEN "11",
            '0' WHEN others;
END a;
```

The majority vote circuit of Figure 5.1 can be described by the truth table in Table 5.2.

Table 5.2 Majority Vote Truth Table

A	*B*	*C*	*Y*
0	0	0	0
0	0	1	0
0	1	0	0
0	1	1	1
1	0	0	0
1	0	1	1
1	1	0	1
1	1	1	1

Since the inputs have different names, rather than the same name with different subscripts, we must group them into an internal 3-bit signal if we want to use them as selection inputs in a selected signal assignment statement. We do this by defining a signal called **inputs** (we can choose any name for this signal) in the declaration region of the architecture body, just before the BEGIN statement:

```
SIGNAL inputs: STD_LOGIC_VECTOR(2 downto 0);
```

A signal is an internal construct that connects two or more portions of a design, much like a piece of wire. Within the architecture body, we assign the inputs a, b, and c to the various bits of the signal. We can do this one of two ways. We can concatenate (link in sequence) the input bits, using the & operator:

```
inputs <= a & b & c;
```

Or, we can assign each input bit explicitly to the bits of the signal:

```
inputs(2) <= a;
inputs(1) <= b;
inputs(0) <= c;
```

Figures 5.2 and 5.3 show the effect of assigning the input ports to the internal 3-bit signal. Figure 5.2 shows the signal as a 3-bit block, with the bits designated from 2 on the left to 0 on the right, with the input port names assigned to the bit positions of the signal. Figure 5.3 shows how the three input ports can be imagined as being bundled together in one signal within the VHDL design entity.

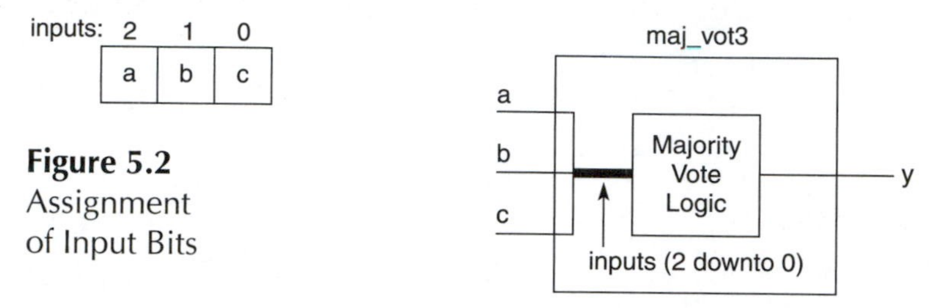

Figure 5.2 Assignment of Input Bits

Figure 5.3 Mapping of Input Ports to a 3-bit Signal

The following code shows how the majority vote circuit is encoded as a truth table:

```
-- maj_vot3.vhd
-- Majority vote circuit using a selected signal assignment statement
LIBRARY ieee;
USE ieee.std_logic_1164.ALL;

ENTITY maj_vot3 IS
PORT(
   a, b, c : IN    STD_LOGIC;
   y       : OUT   STD_LOGIC);
END maj_vot3;

ARCHITECTURE a OF maj_vot3 IS
   SIGNAL inputs : STD_LOGIC_VECTOR(2 downto 0);
BEGIN
   inputs <= a&b&c;

   WITH inputs SELECT
       y  <=  '0' WHEN "000",
              '0' WHEN "001",
              '0' WHEN "010",
              '1' WHEN "011",
              '0' WHEN "100",
              '1' WHEN "101",
              '1' WHEN "110",
              '1' WHEN "111",
              '0' WHEN others;
END a;
```

Procedure

1. Use the MAX+PLUS II Text Editor to enter the VHDL file for the majority vote circuit, using a concurrent signal assignment statement, as listed in the Experimental Notes for this lab exercise. Save the file as *drive:***max2work\lab05\maj_vot2.vhd** and set the project to the current file. Assign the target device as EPM7128SLC84-7 or EPM7128SLC84-15, as appropriate for your board. Assign the pin numbers in Table 5.3 to the circuit inputs and outputs.

Table 5.3 Pin Numbers for the Majority Vote Circuit

Pin Name	Pin Number
A	34
B	33
C	36
Y	44

Compile the project and download it to the Altera UP-1 or RSR PLDT-2 board. Connect three DIP switches to inputs A, B, and C and an LED indicator to output Y. (The RSR board requires no wiring provided the DIP switch and LED jumpers on HD1 and HD2 are installed.) Take the truth table of the circuit and show it to your instructor.

Truth Table:

Instructor's Initials: __________

Note A major difference between the Altera UP-1 board and the RSR PLDT-2 board is that the LEDs and numerical displays on the **Altera UP-1 board** are **active-LOW** (illuminated by logic 0) and the LEDs and numerical displays on the **RSR PLDT-2** board are **active-HIGH** (illuminated by logic 1). If you are using the Altera UP-1 board, you must remember that an ON LED represents a 0 or you will have to invert all circuit outputs to make ON LEDs indicate logic HIGH levels.

2. Repeat procedure 1, but encode the majority vote circuit using a selected signal assignment statement, as shown in the Experimental Notes section of this lab exercise. Save the file as ***drive:*\max2work\lab05\maj_vot3.vhd** and set the project to the current file. Assign the pin numbers as in Table 5.3 and compile the design. Take the truth table of the programmed circuit and show it to your instructor.

   Truth Table:

Instructor's Initials: __________

3. Refer to the truth table shown in Table 5.4. Write the Boolean expression represented by the truth table and simplify it as much as possible.

Table 5.4 Truth Table for Procedures 3 to 5

A	*B*	*C*	*D*	*Y*
0	0	0	0	1
0	0	0	1	1
0	0	1	0	0
0	0	1	1	0
0	1	0	0	0
0	1	0	1	0
0	1	1	0	1
0	1	1	1	0
1	0	0	0	0
1	0	0	1	0
1	0	1	0	0
1	0	1	1	1
1	1	0	0	0
1	1	0	1	1
1	1	1	0	0
1	1	1	1	0

Boolean expression:

Instructor's Initials: ________

4. Write a VHDL design file that uses a concurrent signal assignment statement to encode the Boolean expression derived in procedure 3. Save the file as *drive:*\max2work\lab05\proc3.vhd.

   Note that in VHDL all logical operations have equal precedence, so precedence must be explicitly stated with parentheses. For example, the product term $\overline{A}\, B\, C\, \overline{D}$ must be written as `((not a) and b and c and (not d))`.

   Assign the target device as EPM7128SLC84-7 or EPM7128SLC84-15, as appropriate, and the pins according to Table 5.5. Compile the file and download it to the Altera UP-1 or RSR PLDT-2 board. Connect DIP switches to A, B, C, and D and an LED indicator to Y. (The RSR board is already wired via the jumpers on HD1 and HD2.) Take the truth table of the programmed circuit and show your instructor.

Table 5.5 Pin Assignments for the Circuit Described in Table 5.4

Pin Name	Pin Number
A	34
B	33
C	36
D	35
Y	44

Truth Table:

Instructor's Initials: ________

5. Write a VHDL file that uses a selected signal assignment statement to encode the truth table of Table 5.4. Save the file as ***drive:*\max2work\lab05\proc5.vhd** and set the project to the current file. Assign the target device and pin assignments as in procedure 4. Compile the file and download it to the CPLD board. Take the truth table of the programmed device and show your instructor.

Truth Table:

Instructor's Initials: __________

Assignment Questions

1. Based on the programs written in procedures 4 and 5, state some advantages and disadvantages of describing a combinational logic circuit with a concurrent signal assignment statement and with a selected signal assignment statement. What types of circuits lend themselves best to each method?

2. Complete problems 4.15 through 4.20, 4.22, and 4.26 in the textbook. Note that the circuits in problems 4.22 and 4.26 are best done by using different methods studied in this lab.

Lab 6

Binary and Seven-Segment Decoders

Name ______________________ Class __________ Date __________

Objectives Upon completion of this laboratory exercise, you should be able to:

- Enter the design for a binary decoder in MAX+PLUS II as a Graphic Design File or a VHDL design entity.
- Create a MAX+PLUS II simulation of a binary decoder.
- Use VHDL to create a seven-segment decoder.
- Test the binary and seven-segment decoders on the Altera UP-1 board or equivalent.

Reference Dueck, Robert K., *Digital Design with CPLD Applications and VHDL*

Chapter 5: Combinational Logic Functions
5.1 Decoders

Equipment Required CPLD Trainer:
Altera UP-1 Circuit Board with ByteBlaster Download Cable, or
RSR PLDT-2 Circuit Board with Straight-Through Parallel Port Cable, or
Equivalent CPLD Trainer Board with Altera EPM7128S CPLD
MAX+PLUS II Student Edition Software
AC Adapter, minimum output: 7 VDC, 250 mA DC
Anti-static wrist strap
#22 solid-core wire
Wire strippers

Experimental Notes

Binary Decoder

A decoder is a combinational circuit with one or more outputs, each of which activates in response to a unique binary input value. For example, a 2-line-to-4-line decoder, shown in Figure 6.1, has two inputs, D_1 and D_0, and four outputs, Y_0, Y_1, Y_2, and Y_3. Y_0 is active when $D_1D_0 = 00$, Y_1 activates when $D_1D_0 = 01$, Y_2 is active for $D_1D_0 = 10$, and Y_3 activates for $D_1D_0 = 11$. Only one output is active at any time.

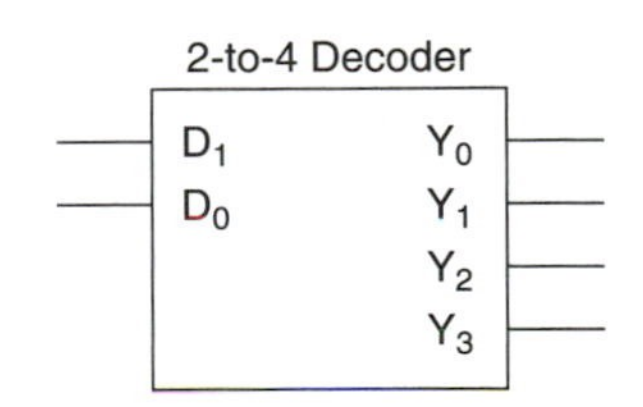

Figure 6.1 2-Line-to-4-Line Decoder

The circuit for the decoder is shown in Figure 6.2. Each AND gate is configured so that its output goes HIGH with a particular value of D_1D_0. In general, the active output is the one whose subscript is equivalent to the binary value of the input. For example, if $D_1D_0 = 10$, only the AND gate for output Y_2 has two HIGH inputs and therefore a HIGH output.

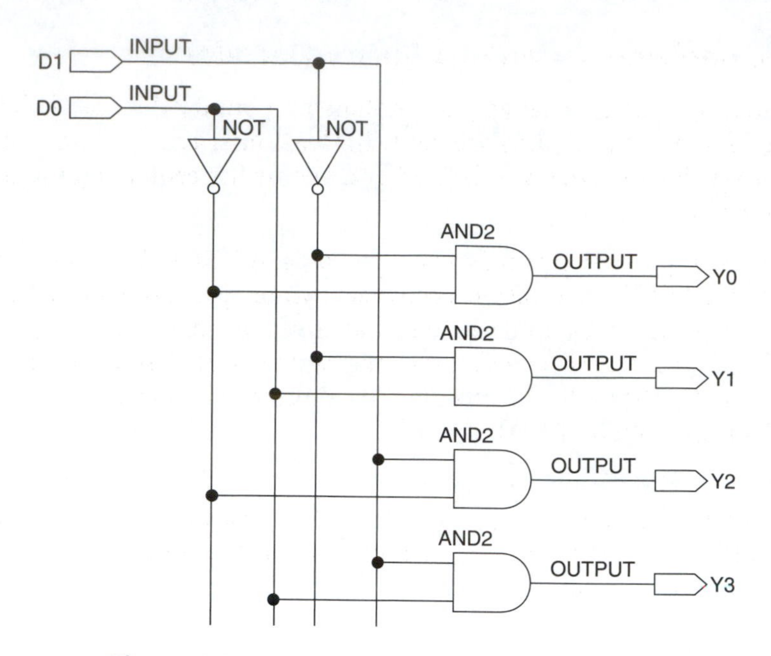

Figure 6.2 2-Line-to-4-Line Decoder Circuit

A standard part of the CPLD design cycle is to create a simulation of a design before programming it into a CPLD. A simulation is a timing diagram which is generated by specifying a set of input waveforms to the design under test. The simulation software examines the design equations and input logic waveforms and calculates the response of the digital circuit, which is displayed as a set of output waveforms. The simulation allows us to determine if the design is working as planned by observing the response of the design to a defined input. A good simulation will test the design under all possible input conditions.

Figure 6.3 shows a simulation of a 2-line-to-4-line decoder, generated in MAX+PLUS II. The inputs D_1 and D_0 are grouped together as a single 2-bit value that can range from 0 to 3 in decimal (equivalent to 00 to 11 in binary). The D inputs have an increasing 2-bit binary count applied to them: the inputs start at 0, increase up to 3, then go back to 0 and repeat. In response, the Y outputs activate one at a time by going HIGH in a sequence that corresponds to the change on the D inputs. A procedure for creating a simulation is shown in *Digital Design with CPLD Applications and VHDL* on pages 162–167.

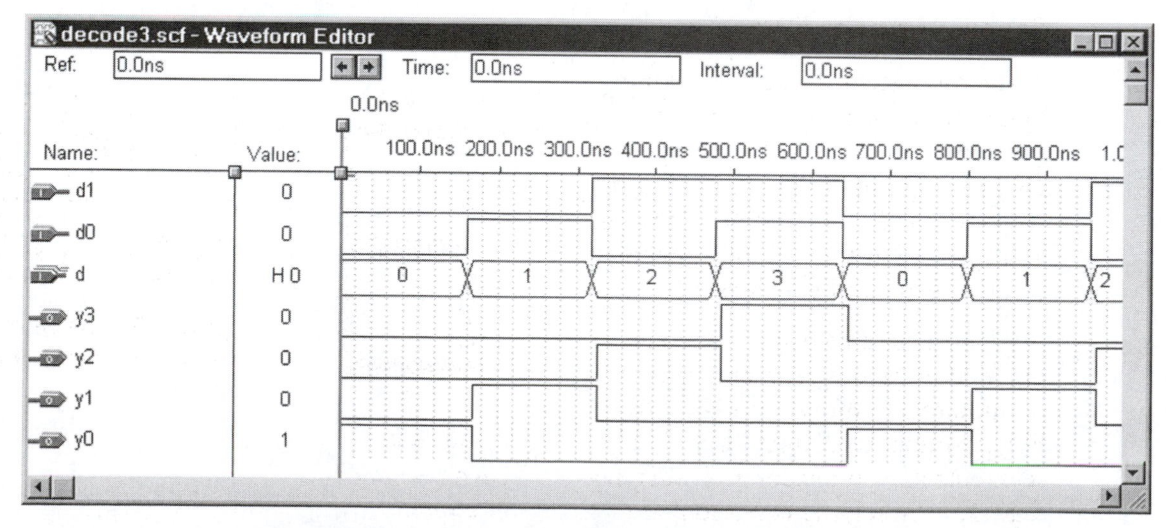

Figure 6.3 Simulation of a 2-Line-to-4-Line Decoder

VHDL Implementation of a Binary Decoder

The most straightforward way to implement a binary decoder in VHDL is to use a selected signal assignment statement. In this construct, the four outputs of a 2-line-to-4-line decoder are defined as a vector for each combination of inputs, also expressed as a vector.

For example, for inputs $D_1D_0 = 01$, output Y_1 will be HIGH and the remaining outputs are LOW. The output vector is written as 0010, where the leftmost bit is Y_3 and the rightmost bit is Y_0, as defined in the entity declaration. (The output vector could also be defined with Y_0 on the left, by changing the port definition to read `y : OUT STD_LOGIC_VECTOR (0 to 3)`. In this case, the output vector for input 01 is 0100, since Y_1 is now second from the left.)

The following VHDL design entity illustrates the use of a selected signal assignment to define a decoder. The **others** clause is required since the STD_LOGIC type encompasses values other than 0 and 1. The default value in the **others** clause shows the outputs all inactive (0000) for any unspecified input value.

```
LIBRARY ieee;
USE ieee.std_logic_1164.ALL;

ENTITY dcd2to4 IS
   PORT(
       d  : IN     STD_LOGIC_VECTOR (1 downto 0);
       y  : OUT    STD_LOGIC_VECTOR (3 downto 0));
END dcd2to4;

ARCHITECTURE decoder OF dcd2to4 IS
BEGIN
   WITH d SELECT
       y <=    "0001" WHEN "00",
               "0010" WHEN "01",
               "0100" WHEN "10",
               "1000" WHEN "11",
               "0000" WHEN others;
END decoder;
```

Seven-Segment Decoder

A seven-segment display, shown in Figure 6.4, consists of seven luminous segments, such as LEDs, arranged in a figure-8 pattern. The segments are conventionally designated *a* through *g*, beginning at the top and moving clockwise around the display.

When used to display decimal digits, the various segments are illuminated as shown in Figure 6.5. For example, to display digit 0, all segments are on except *g*. To display digit 1, only segments *b* and *c* are illuminated.

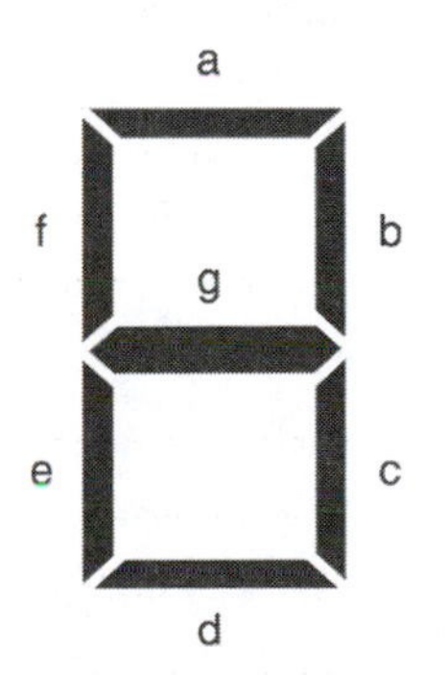

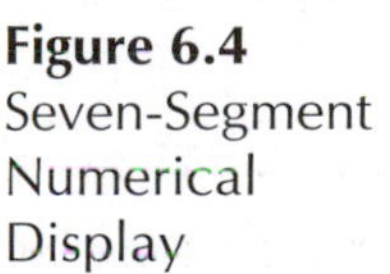
Figure 6.4 Seven-Segment Numerical Display

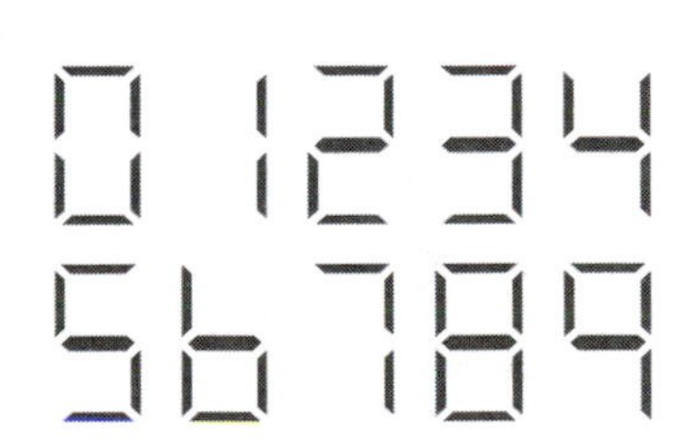

Figure 6.5 Convention for Displaying Decimal Digits

The seven-segment displays on the Altera UP-1 board are configured as common-anode, meaning that the anodes of all LEDs are tied together and connected to the board power supply, V_{CC}. (Refer to Figure 5.24, p. 172 in *Digital Design with CPLD Applications and VHDL.*) To turn on a segment, the cathode end of the LED is set to logic 0 through a current-limiting series resistor. This is illustrated for digits 0 and 1 in the partial truth table shown in Table 6.1.

Table 6.1 Partial Truth Table for a Common Anode BCD-to-7-Segment Decoder

D_3	D_2	D_1	D_0	a	b	c	d	e	f	g
0	0	0	0	0	0	0	0	0	0	1
0	0	0	1	1	0	0	1	1	1	1

The seven-segment displays on the RSR PLDT-2 circuit board are common cathode, or active HIGH. The cathodes of all segment LEDs are tied to ground and each individual segment is illuminated by a HIGH applied to the segment anode through a series resistor. Table 6.2 illustrates a partial truth table for a common cathode decoder, showing the output values for digits 0 and 1.

Table 6.2 Partial Truth Table for a Common Cathode BCD-to-7-Segment Decoder

D_3	D_2	D_1	D_0	a	b	c	d	e	f	g
0	0	0	0	1	1	1	1	1	1	0
0	0	0	1	0	1	1	0	0	0	0

The seven-segment displays and series resistors on both the Altera UP-1 board and the RSR PLDT-2 board are hardwired to the board's CPLD. Thus, all that is required to turn on the illuminated segments for each digit is to make the appropriate CPLD pins LOW (Altera UP-1) or HIGH (RSR PLDT-2). A VHDL file for a common anode BCD-to-seven-segment decoder is as follows:

```
-- bcd_7seg.vhd
-- Common Anode BCD-to-seven-segment decoder
LIBRARY ieee;
USE ieee.std_logic_1164.ALL;

ENTITY bcd_7seg IS
   PORT(
        d3, d2, d1, d0      : IN    STD_LOGIC;
        a, b, c, d, e, f, g   : OUT   STD_LOGIC);
END bcd_7seg;

ARCHITECTURE seven_segment OF bcd_7seg IS
   SIGNAL input  : STD_LOGIC_VECTOR (3 downto 0);
   SIGNAL output : STD_LOGIC_VECTOR (6 downto 0);
BEGIN
   input <= d3 & d2 & d1 & d0;
```

```
WITH input SELECT
    output <=  "0000001" WHEN "0000", -- display 0
               "1001111" WHEN "0001", -- display 1
               "0010010" WHEN "0010", -- display 2
               "0000110" WHEN "0011", -- display 3
               "1001100" WHEN "0100", -- display 4
               "0100100" WHEN "0101", -- display 5
               "1100000" WHEN "0110", -- display 6
               "0001111" WHEN "0111", -- display 7
               "0000000" WHEN "1000", -- display 8
               "0001100" WHEN "1001", -- display 9
               "1111111" WHEN others;

- - Separate the output vector to make individual pin outputs.
     a  <=  output(6);
     b  <=  output(5);
     c  <=  output(4);
     d  <=  output(3);
     e  <=  output(2);
     f  <=  output(1);
     g  <=  output(0);

END seven_segment;
```

Procedure

Note In the following procedures, there are two sets of instructions: one set for the Altera UP-1 board, whose LEDs and seven-segment displays are active-LOW and another set for the RSR PLDT-2 board, whose LEDs and seven-segment displays are active-HIGH. In places where there is no difference in procedure for the two boards, there is only one set of instructions. Where there is a difference, follow only those instructions that apply to your board.

Binary Decoder (Graphic Design File; Altera UP-1 Board)

1. Use the Graphic Editor in MAX+PLUS II to create a 3-line-to-8-line decoder with active-LOW outputs. Designate the inputs D2, D1, and D0 and the outputs Y0 through Y7. Save the file as *drive:***max2work\lab06\3to8dcdl.gdf** and set the project to the current file.

2. Compile the file and follow the procedure below to create a simulation to show that the design is correct.

Binary Decoder (Graphic Design File; RSR PLDT-2 Board)

1. Use the Graphic Editor in MAX+PLUS II to create a 3-line-to-8-line decoder with active-HIGH outputs. Designate the inputs D2, D1, and D0 and the outputs Y0 through Y7. Save the file as *drive:***max2work\lab06\3to8dcdh.gdf** and set the project to the current file.

2. Compile the file and follow the procedure below to create a simulation to show that the design is correct.

Creating a Simulation (Altera UP-1 and RSR PLDT-2)

1. From the MAX+PLUS II **File** menu, select **New**. On the resultant dialog box, select **Waveform Editor File**, with a default file extension **scf.** From the **File** menu, **Save As**

   ***drive:*\max2work\lab06\3to8dcdl.scf** (for the Altera UP-1)
   or
   ***drive:*\max2work\lab06\3to8dcdh.scf** (for the RSR PLDT-2).

2. We specify the inputs and outputs we want to view by selecting **Enter Nodes from SNF** on the **Node** menu, shown in Figure 6.6. In the dialog box that pops up (Figure 6.7), there are two boxes labeled **Available Nodes & Groups** and **Selected Nodes & Groups**, with an arrow **(=>)** pointing from one to the other. Select the **List** button to show the "available" signals and click the arrow to transfer them all to the "selected" box. Click **OK** to close the box.

Figure 6.6 Node Menu

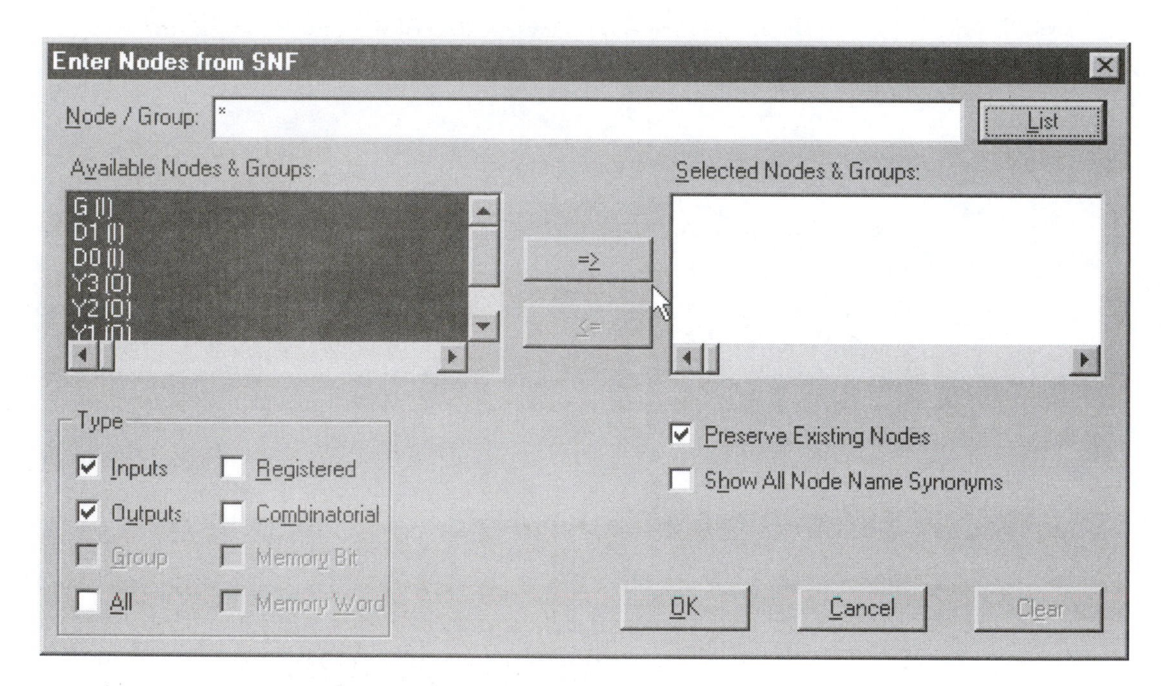

Figure 6.7 Selecting Nodes for Simulation

Figure 6.8 on page 49 shows the simulation waveforms in their uninitialized (default) states. Inputs and outputs are shown by symbols in front of the signal names. Inputs are at logic 0 and outputs indicated as X or unknown values.

3. We should now set the **Grid Size**, which determines the size of the smallest time division in the simulation. To do so, select **Grid Size** from the **Options** menu, shown in Figure 6.9. In the dialog box of Figure 6.10, enter the value **20ns** and click **OK.** (We will use this value for many of our simulations because it corresponds to one half period of the 25.175-MHz oscillator on the Altera UP-1 board. In the simulator, one full period requires two grid spaces. For the RSR PLDT-2 board, we can use a grid of 125 ns to correspond to one half period of the 4-MHz on-board oscillator.)

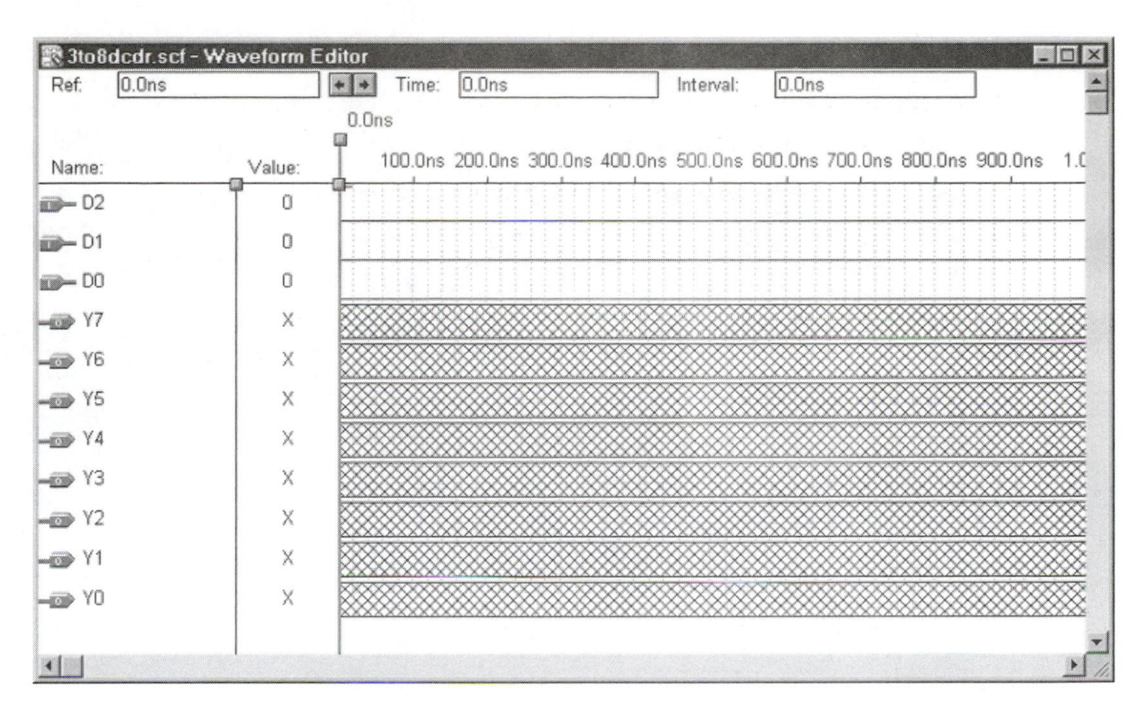

Figure 6.8 Default Waveforms for a Simulation

Figure 6.9 Options Menu

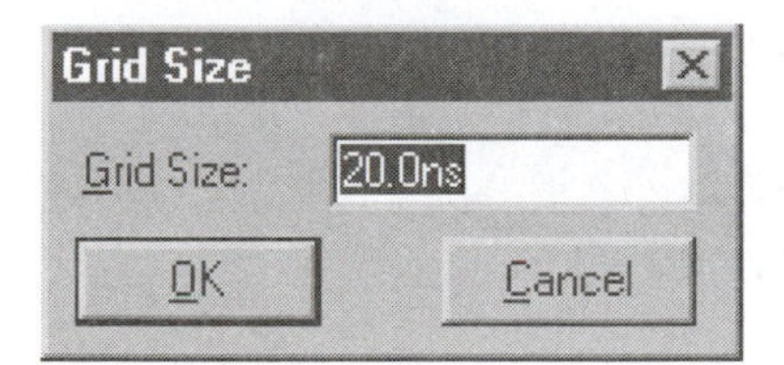

Figure 6.10 Setting the Grid Size for a Simulation

4. When we created the simulation file, the D inputs were entered as separate waveforms. We can join these waveforms to make a **Group**. Highlight D2, D1, and D0 by clicking on the name for D2 and dragging the mouse to D0, as in Figure 6.11 on page 50. (Do not click on the input symbol, as this will move the waveform to another position.) From the **Node** menu or the pop-up menu (accessed by a right-click on the highlighted group), select **Enter Group**, as shown in Figure 6.12 on page 50. The dialog box shown in Figure 6.13 appears, containing the most likely name derived from the highlighted group. Either type a new group name or accept the original name by clicking **OK**.

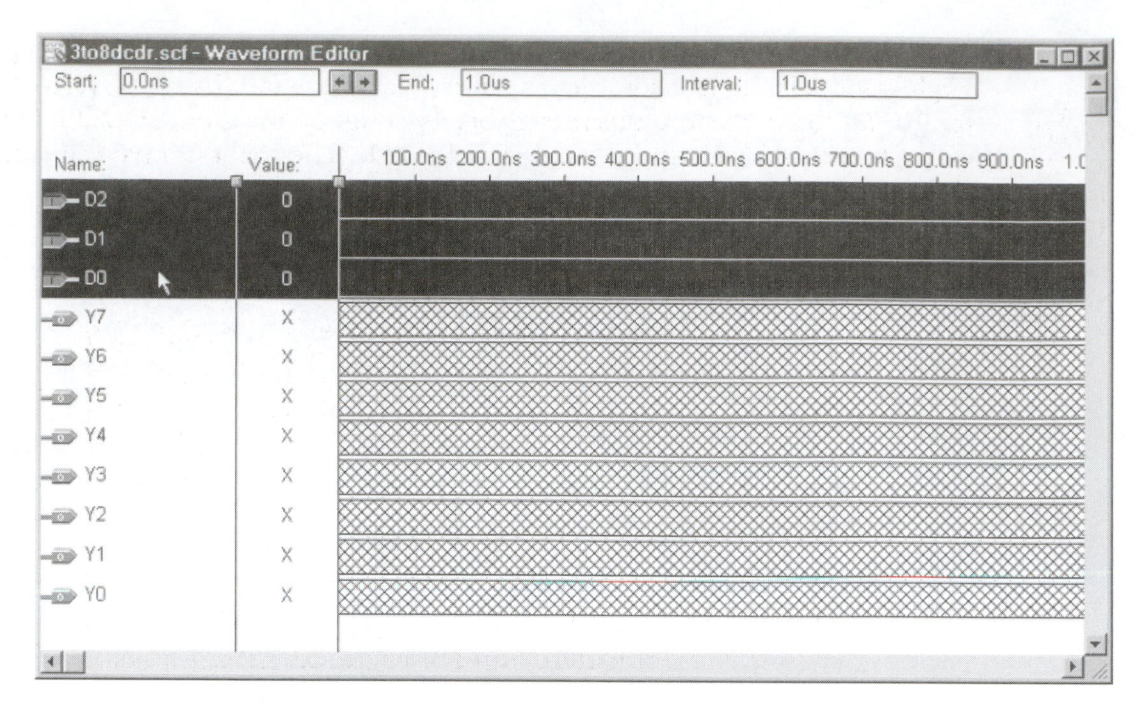

Figure 6.11 Highlighting a Group of Waveforms

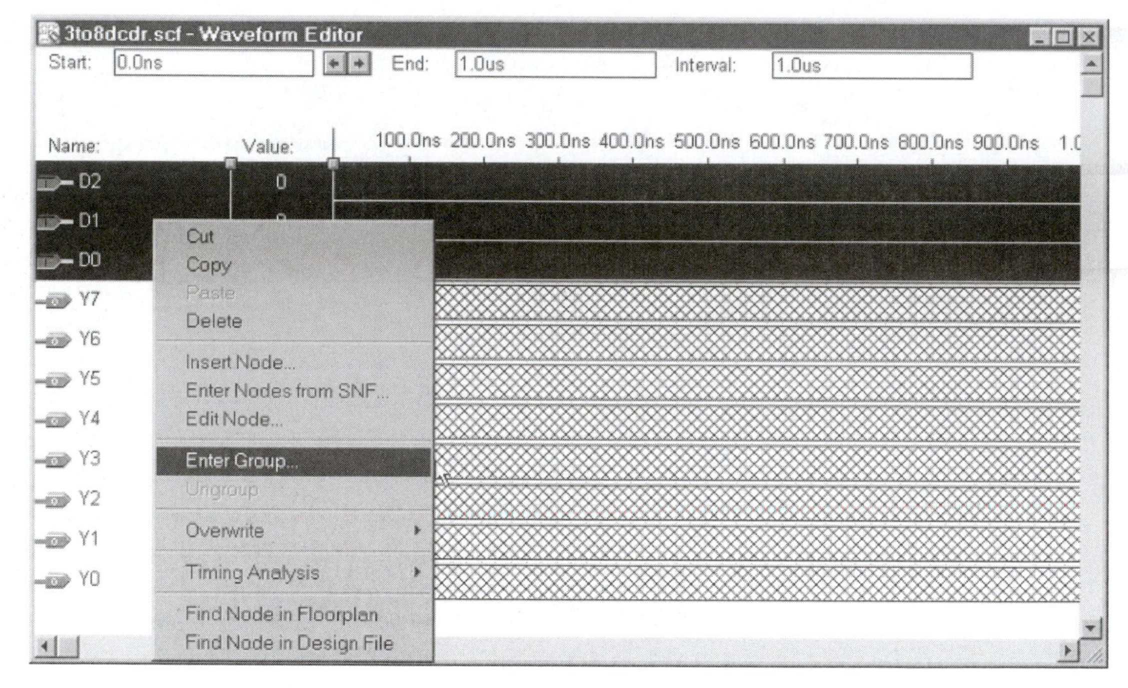

Figure 6.12 "Enter Group" from a Pop-up Menu

Figure 6.13 Enter Group Dialog

Enter Group

Group Name: D[2..0]

Display Gray Code Count As Binary Count

Radix

BIN DEC

OCT HEX

OK

Cancel

5. As a decoder stimulus, we will define an increasing binary count on the D inputs. Highlight the input group by clicking in the **Value** column. Use the **Overwrite Count** toolbar button to create an increasing binary count on the group, D[2..0]. Fill in the dialog box as shown in Figure 6.14 and click **OK**. The count is increased every 40 ns (2×20 ns), as shown in Figure 6.15.

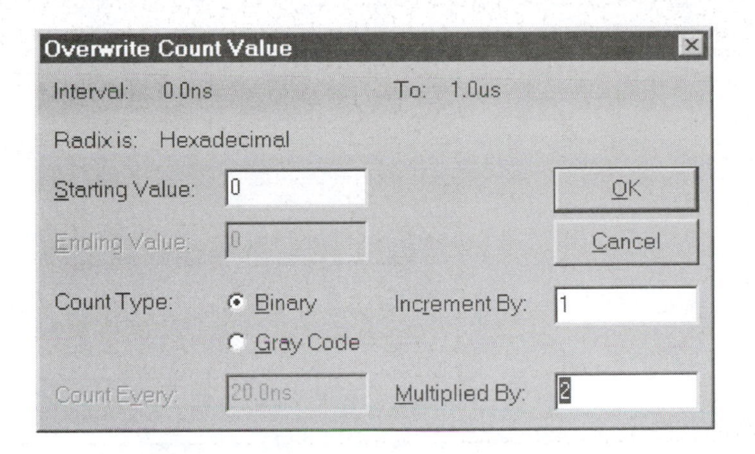

Figure 6.14 Overwrite Count Dialog

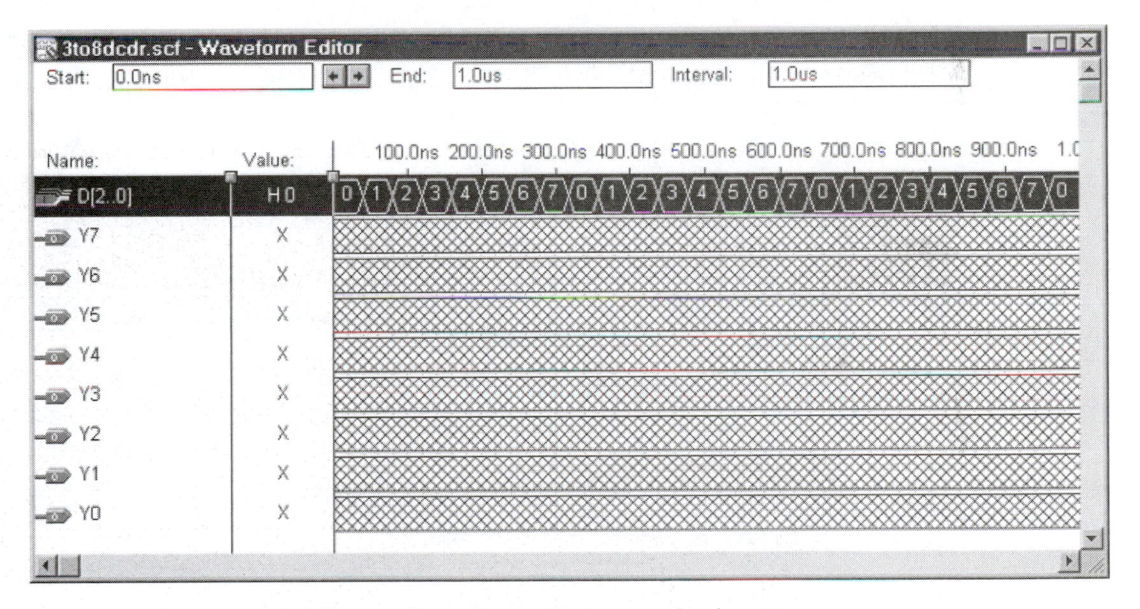

Figure 6.15 Count Value Applied to Group

6. Save the file. From the **MAX+PLUS II** menu, bring the **Simulator** to the front and click **Start.** When the simulation is finished (almost immediately), click **Open SCF** and maximize the window. From the **View** menu, select **Fit in Window** or select the toolbar button for this function.

   The simulator output, shown in Figure 6.16 for the active-LOW decoder (Altera UP-1) and Figure 6.17 for the active-HIGH decoder (RSR PLDT-2), shows the result of a repeating binary count at the decoder input. The outputs activate in a repeating sequence, from Y0 to Y7. Show the simulation to your instructor.

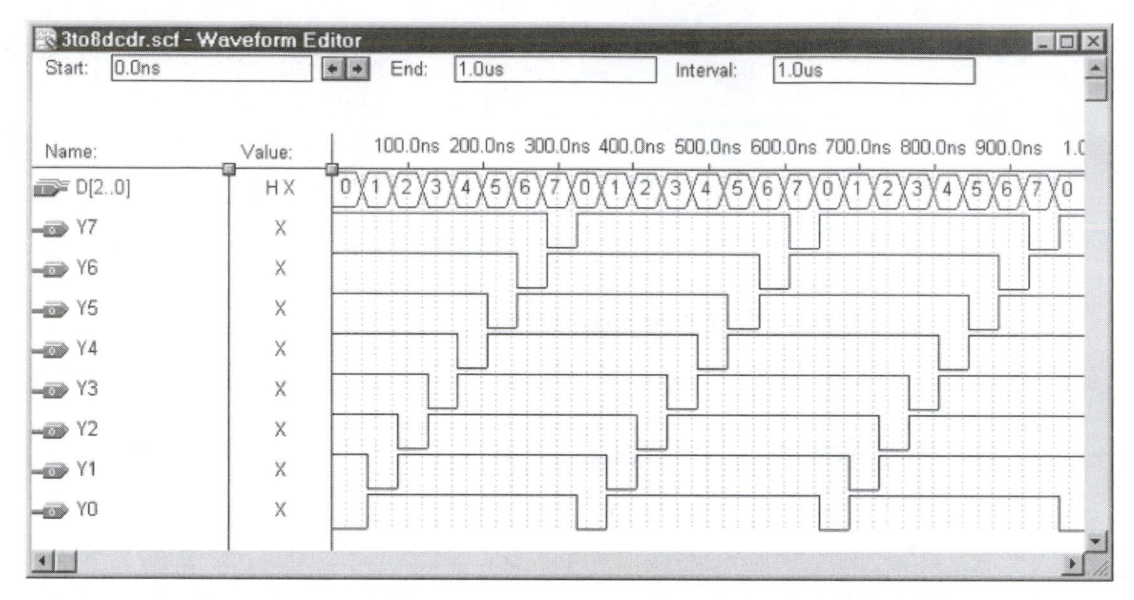

Figure 6.16 Simulation of an Active-LOW Decoder (Altera UP-1)

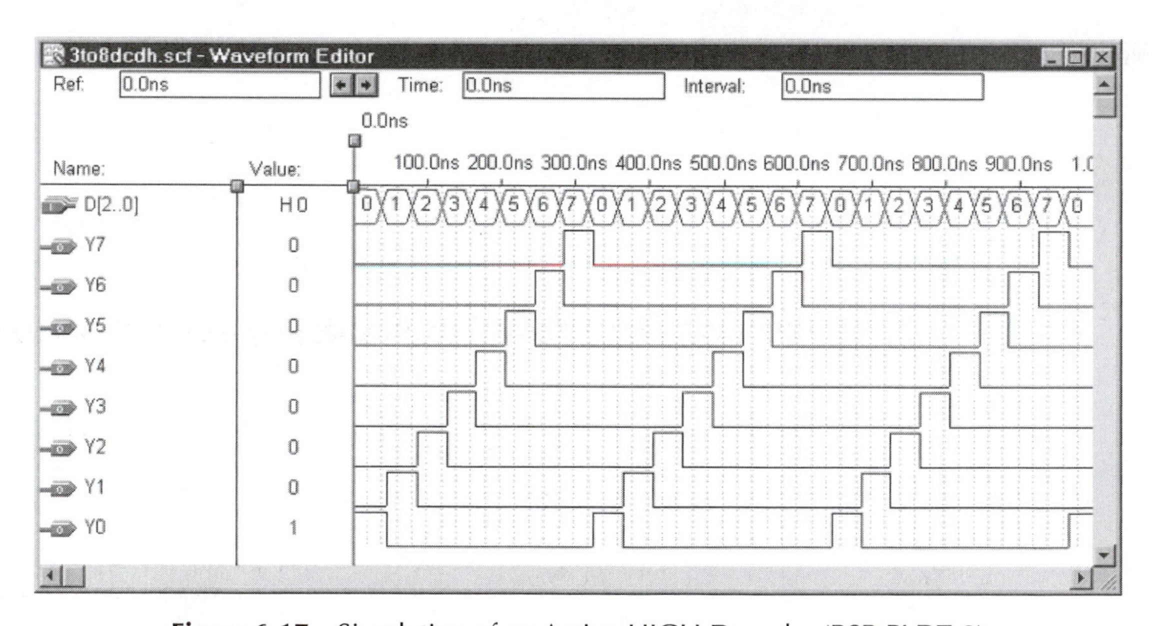

Figure 6.17 Simulation of an Active-HIGH Decoder (RSR PLDT-2)

Instructor's Initials: ________

Changing the Timing Length of the Simulation

The default value of the simulation timing length is 1 μs, written **1.0us**. If our design requires a longer or shorter simulation time, we can set the length as follows. (The example is shown for the active-LOW decoder, but will work similarly for the active-HIGH decoder.)

1. Select **End Time** (**File** menu, Figure 6.18) and enter **1.5us** for the new time for the end of simulation in the dialog box of Figure 6.19. The notation **1.5us** represents a time of 1.5 μs.

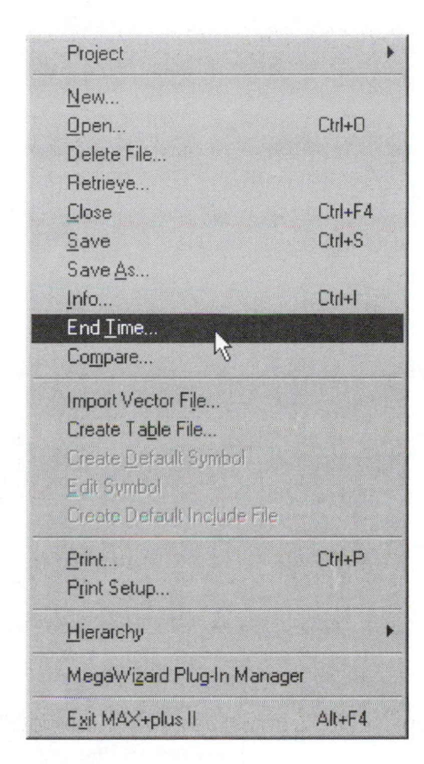

Figure 6.18 File Menu (End Time)

End Time
Time: 1.5us
OK
Cancel

Figure 6.19 End Time Dialog

2. Since the new end time is longer than the previous value, it adds blank space to the end of the simulation, as shown in Figure 6.20. Zoom back to see the entire waveform.

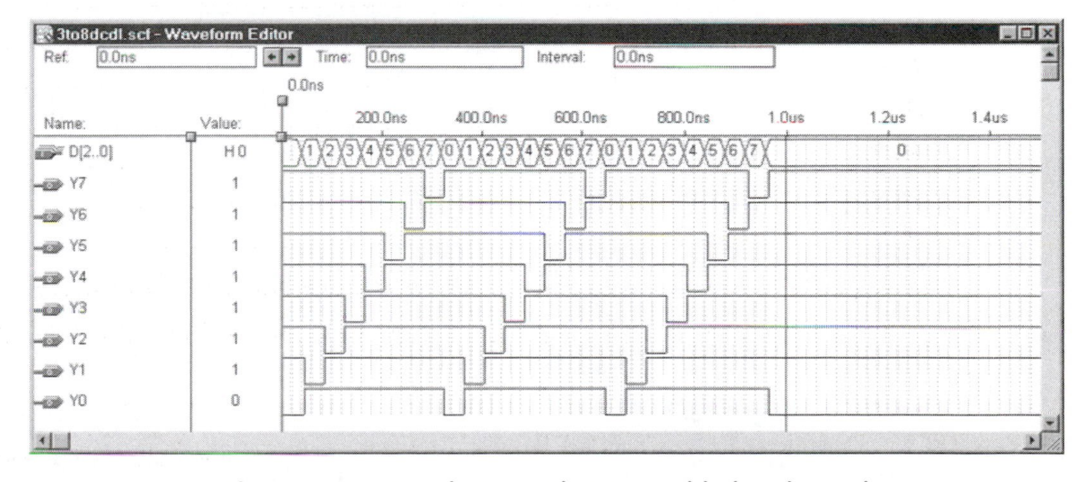

Figure 6.20 Simulation with 0.5μs Added to the end

3. Highlight the D input waveform and apply an increasing binary count, as shown in Figure 6.21. Run the simulation again to get the waveforms in Figure 6.22 (Altera UP-1) or Figure 6.23 (RSR PLDT-2). Show the simulation to your instructor.

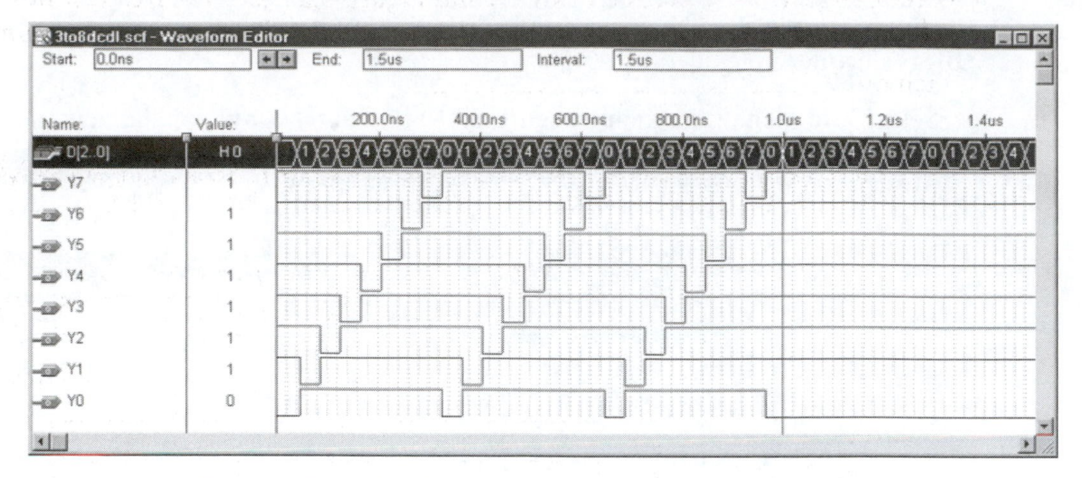

Figure 6.21 Simulation with Rewritten Count Waveform

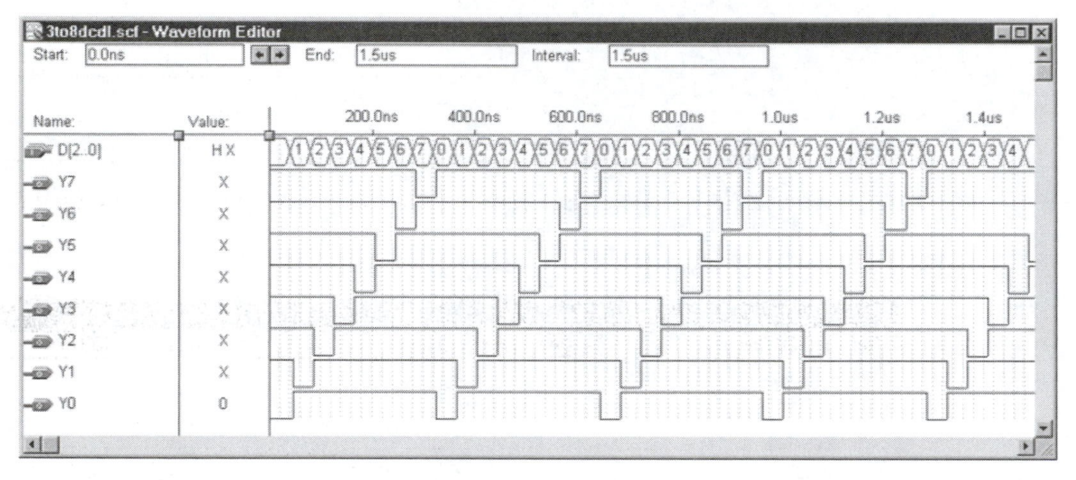

Figure 6.22 New Simulation Waveform (Active-LOW Decoder)

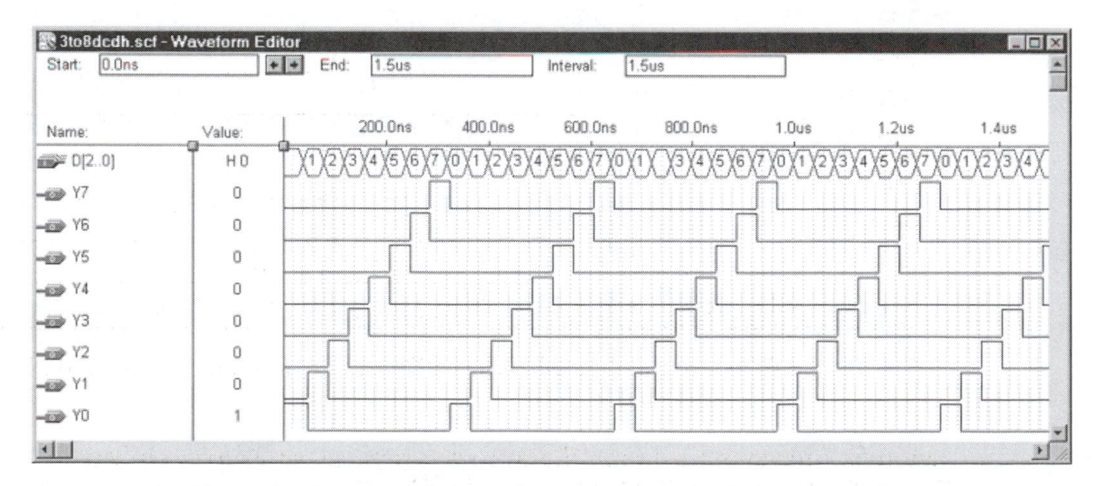

Figure 6.23 New Simulation Waveform (Active-HIGH Decoder)

Instructor's Initials: ________

Simulation as a Diagnostic Tool

Simulations are very useful for detecting design errors. For example, the simulations in Figure 6.24 (active-LOW; Altera UP-1) and Figure 6.25 (active-HIGH; RSR PLDT-2) each show that one of the logic gates in the decoder has been incorrectly connected.

Which one? ______________________________

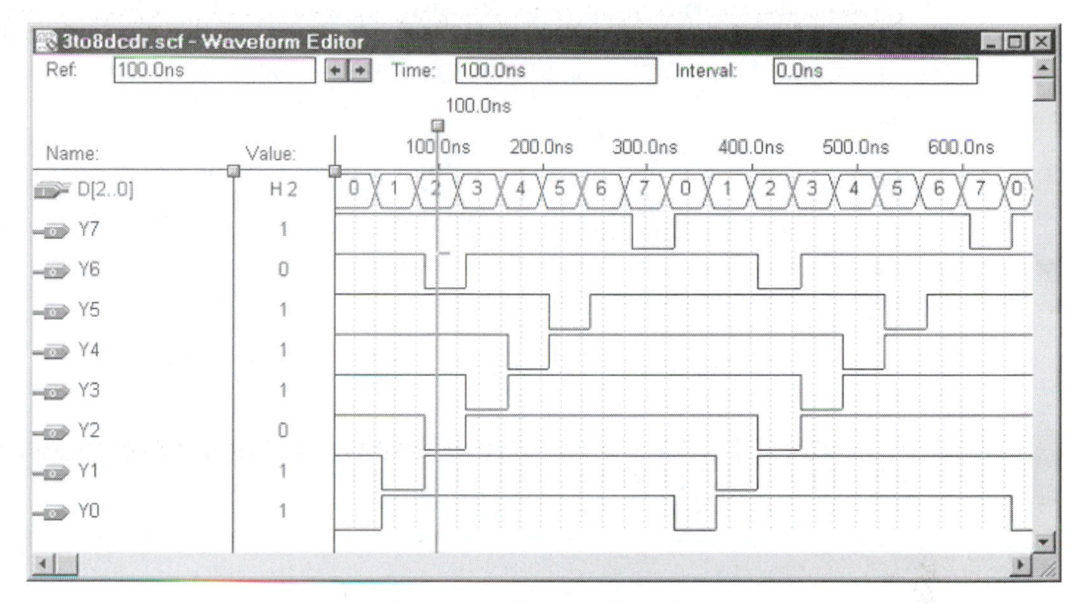

Figure 6.24 Simulation Showing a Design Error in an Active-LOW Decoder

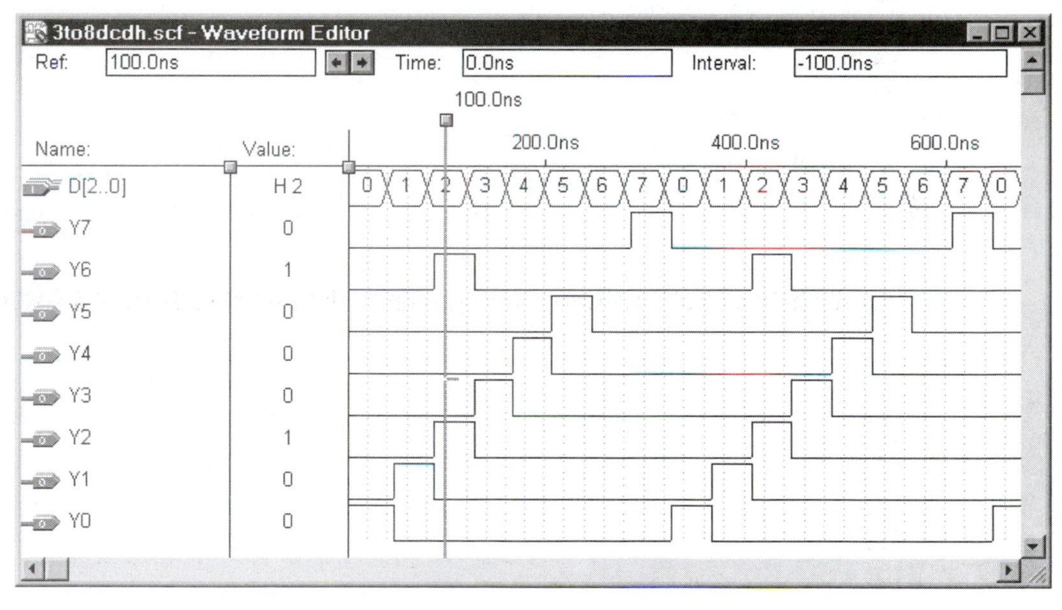

Figure 6.25 Simulation Showing a Design Error in an Active-HIGH Decoder

Modify your graphic design file so that it replicates the error shown in the simulation of Figure 6.24 or Figure 6.25. Compile the file and run the simulation again. Show the revised **gdf** and simulation to your instructor.

Instructor's Initials: __________

Programming and Testing the Binary Decoder

Restore the **gdf** file of the decoder to its original configuration (without the design error) and assign pins to the decoder, as listed in Table 6.3. Compile the design again and run the simulation again to verify correct operation. Download the decoder design to the CPLD board. If using the Altera UP-1 board, connect three DIP switches to the pins assigned to the D inputs and eight LEDs to the decoder outputs. (The RSR PLDT-2 board is already configured with jumpers.) Demonstrate the operation of the decoder to your instructor and take its truth table, recalling that the LEDs on the Altera UP-1 board are active-LOW and the LEDs on the RSR PLDT-2 board are active-HIGH.

Table 6.3 Pin Assignments for a 3-Line-to-8-Line Decoder

Pin Name	Pin Number	Device
D2	34	SW1-1
D1	33	SW1-2
D0	36	SW1-3
Y0	44	LED1
Y1	45	LED2
Y2	46	LED3
Y3	48	LED4
Y4	49	LED5
Y5	50	LED6
Y6	51	LED7
Y7	52	LED8

Truth Table:

Instructor's Initials: __________

VHDL Implementation of a Binary Decoder

1. Write a VHDL file for a 3-line-to-8-line decoder with active-LOW outputs. Use a selected signal assignment statement. Save the file as ***drive:*\max2work\lab06\dcd3to8l.vhd** and set the project to the current file. Compile the design and create a simulation for the decoder to verify its correct operation. Show the simulation to your instructor.

   Instructor's Initials: ________

2. Assign pin numbers to the VHDL decoder, as shown in Table 6.3. Compile the design again and download it to the Altera UP-1 or RSR PLDT-2 board. Demonstrate the decoder operation to your instructor.

   Instructor's Initials: ________

3. Modify the VHDL decoder to make its outputs active-HIGH. Save the file as ***drive:*\max2work\lab06\dcd3to8h.vhd** and set the project to the current file. Compile the file and simulate its operation. (**Tip:** Save time by saving the simulation for the decoder with active-LOW outputs under the name for the active-HIGH design and run the simulation again.) Show the simulation to instructor.

   Instructor's Initials: ________

4. Assign pin numbers to the active-HIGH decoder.

5. Compile the VHDL file again, download it to the Altera UP-1 or RSR PLDT-2 board, and demonstrate its operation.

   Instructor's Initials: ________

Hexadecimal-to-Seven-Segment Decoder

1. Create a hexadecimal-to-seven-segment decoder in VHDL, using the segment patterns in Figure 6.26 as a model. The decoder for the Altera UP-1 board will be common anode. For the RSR PLDT-2 board, it will be common cathode. Note that the patterns for digits 6 and 9 are shown differently than in Figure 6.5. Save the VHDL file as ***drive:*\max2work\lab06\hex7seg.vhd** and set the project to the current file.

Figure 6.26 Hexadecimal Digit Display Format

2. Assign pin numbers as shown in Table 6.4. Compile the file and download it to the CPLD board. Demonstrate its operation to your instructor.

Table 6.4 Pin Assignments for Hex-to-Seven-Segment Decoder

Pin Name	Pin Number	Device
D3	33	SW1-1
D2	34	SW1-2
D1	36	SW1-3
D0	35	SW1-4
a	69	Seven-segment display
b	70	
c	73	
d	74	
e	76	
f	75	
g	77	

Instructor's Initials: ________

Assignment Questions

Complete problems 5.8, 5.10, and 5.16 in *Digital Design with CPLD Applications and VHDL.*

Lab 7

Multiplexer Applications

Name ______________________ Class __________ Date ______________

Objectives Upon completion of this laboratory exercise, you should be able to:

- Enter the logic circuit of a 4-to-1 multiplexer (MUX) as a Graphic Design File, using Altera's MAX+PLUS II CPLD design software.
- Create a MAX+PLUS II simulation file for the 4-to-1 multiplexer described above.
- Create a hierarchical design in the MAX+PLUS II Graphic Editor that contains a multiplexer and other components.
- Download the 4-to-1 MUX to a CPLD on an Altera UP-1 or RSR PLDT-2 circuit board and test its function.
- Write VHDL design code for an 8-to-1 multiplexer, create a simulation for the design, download and test it.
- Use an 8-bit MUX as a programmable waveform generator.
- Write VHDL code for a quadruple (4-bit-wide) 2-to-1 multiplexer. Simulate and download the file.
- Test the function of the quad 2-to-1 MUX by switching alternate sets of data to a seven-segment numerical decoder.

Reference Dueck, Robert K., *Digital Design with CPLD Applications and VHDL*

Chapter 4: Introduction to PLDs and MAX+PLUS II
4.5 Hierarchical Design

Chapter 5: Combinational Logic Functions
5.3 Multiplexers

Equipment Required

CPLD Trainer:
Altera UP-1 Circuit Board with ByteBlaster Download Cable, or
RSR PLDT-2 Circuit Board with Straight-Through Parallel Port Cable, or
Equivalent CPLD Trainer Board with Altera EPM7128S CPLD
MAX+PLUS II Student Edition Software
AC Adapter, minimum output: 7 VDC, 250 mA DC
Anti-static wrist strap
#22 solid-core wire
Wire strippers

Experimental Notes

A multiplexer (abbreviated MUX) is a device for switching one of several digital signals to an output, under the control of another set of binary inputs. The inputs to be switched are called the **data inputs**; those that determine which signal is directed to the output are called the **select inputs**.

Figure 7.1 shows the logic diagram of a 4-to-1 multiplexer, with data inputs labeled D_3 to D_0 and the select inputs labeled S_1 and S_0. By examining the circuit, we can see that the 4-to-1 MUX is described by the following Boolean equation:

$$Y = D_0\overline{S}_1\overline{S}_0 + D_1\overline{S}_1S_0 + D_2S_1\overline{S}_0 + D_3S_1S_0$$

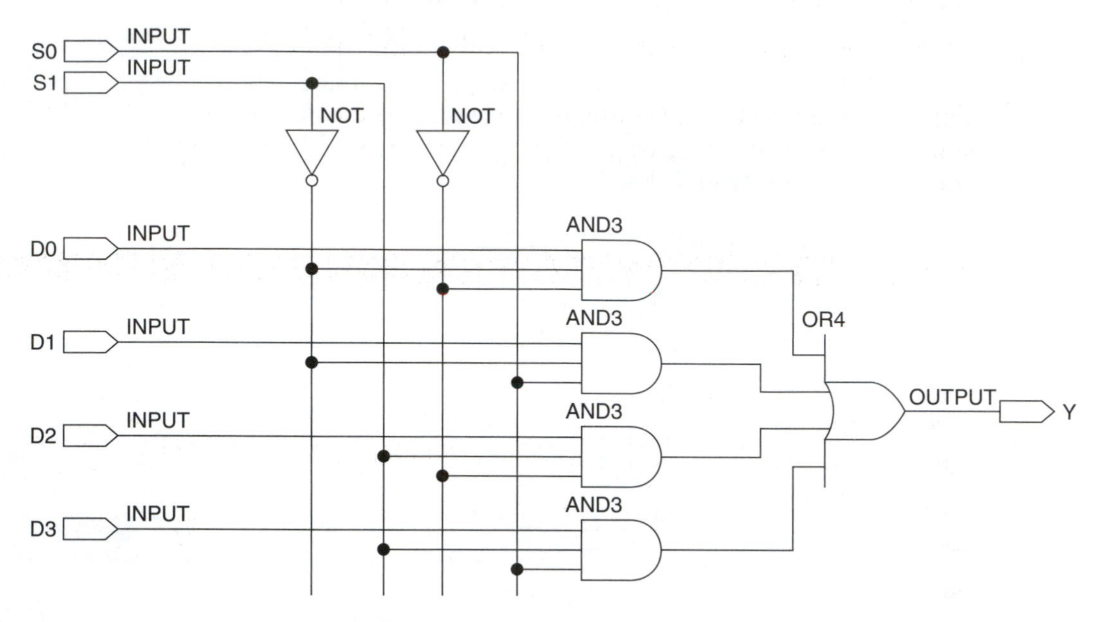

Figure 7.1 4-to-1 Multiplexer

For any given combination of S_1S_0, only one of the above four product terms will be enabled. For example, when $S_1S_0 = 10$, the equation evaluates to:

$$Y = (D_0 \cdot 0) + (D_1 \cdot 0) + (D_2 \cdot 1) + (D_3 \cdot 0) = D_2$$

The MUX equation can be described by a truth table, as in Table 7.1. The subscript of the selected data input is the decimal equivalent of the binary combination S_1S_0.

Table 7.1 **Truth Table of a 4-to-1 MUX**

S_1	S_0	Y
0	0	D_0
0	1	D_1
1	0	D_2
1	1	D_3

Multiplexers can be implemented in MAX+PLUS II as a Graphic Design File similar to Figure 7.1 or as a VHDL design entity. Implementation and applications of multiplexers are described in more detail in *Digital Design with CPLD Applications and VHDL* on pages 185–197.

Multiplexers are very versatile circuits. In this lab exercise, we will use them to switch one of four time-varying waveforms to an output, generate a programmable digital waveform pattern, and switch multibit data to a seven-segment display.

Procedure

Graphic Design File and Simulation for 4-to-1 Multiplexer

1. Create a Graphic Design File for a 4-to-1 multiplexer as shown in Figure 7.1. Save the file as ***drive:*\max2work\lab07\4to1mux.gdf.** (**Tip:** You can place the inverters vertically, as shown in Figure 7.1, by entering the NOT symbol, right-clicking on the symbol, and choosing **Rotate, 270°** from the pop-up menu.)

   Set the project to the current file and compile the design.

2. Figure 7.2 shows a set of simulation waveforms created by the MAX+PLUS II simulation tool for the 4-to-1 MUX. Create the simulation waveforms using the specifications given in Table 7.2.

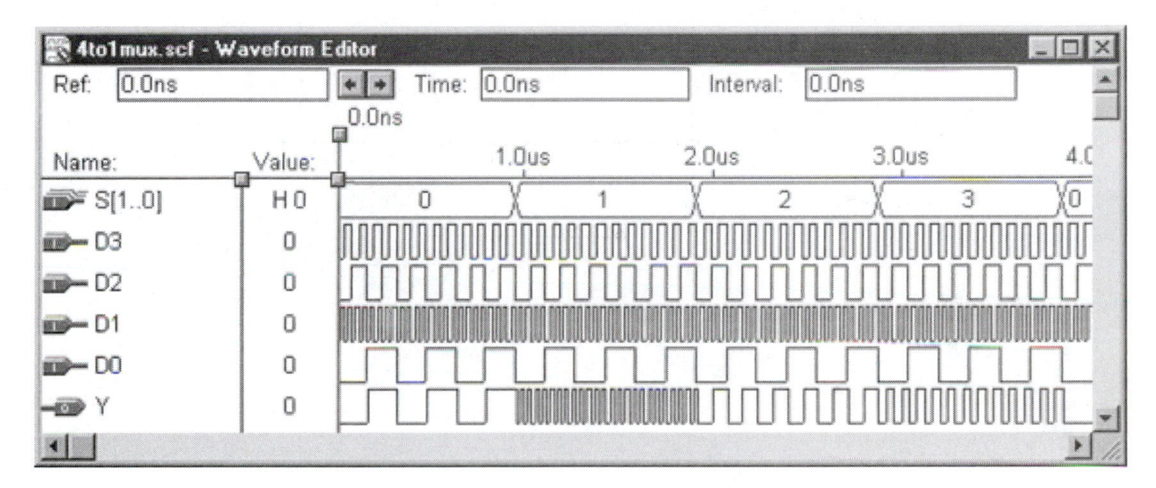

Figure 7.2 Simulation of a 4-to-1 Multiplexer

We will use a set of periodic pulse waveforms (called "clock" waveforms in the simulator) of four different frequencies for the inputs D_3 through D_0: 1, 2, 4, and 8 times a standard base frequency. We can tell which input channel is currently selected by examining the output waveform. The output is the same as the currently selected input waveform.

Table 7.2 Simulation Specs for a 4-to-1 MUX

Parameter	Value
End Time	4 μs
Grid Size	20 ns
Binary Count on S[1..0]	20 ns × 48
Clock period on D3	40 ns × 2
Clock period on D2	40 ns × 4
Clock period on D1	40 ns × 1
Clock period on D0	40 ns × 8

To set the base value of clock period, select **Grid Size** in the **Options** menu and enter **20.0ns**. This is one half the desired base-value clock period, since one clock cycle takes two grid spaces.

Set D_3 to show a pulse waveform by highlighting the D_3 input, as shown in Figure 7.3 (click on the input name). Click on the **Overwrite Clock Waveform** button on the toolbar at the left-hand side of the screen. If we want D_3 to have a frequency 2 times the base value, fill in the dialog box as shown in Figure 7.4 and click **OK**. The resulting waveform is shown in Figure 7.5.

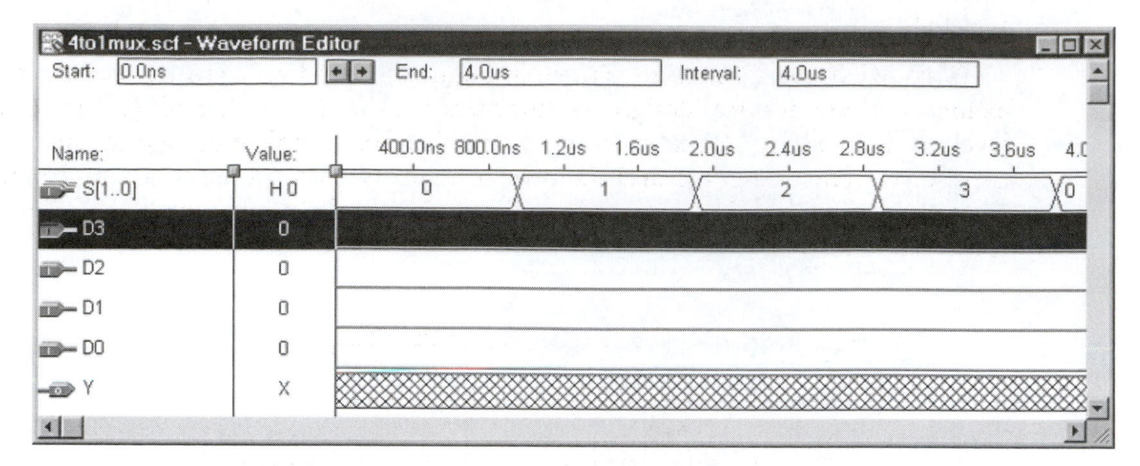

Figure 7.3 Highlighting an Input Line

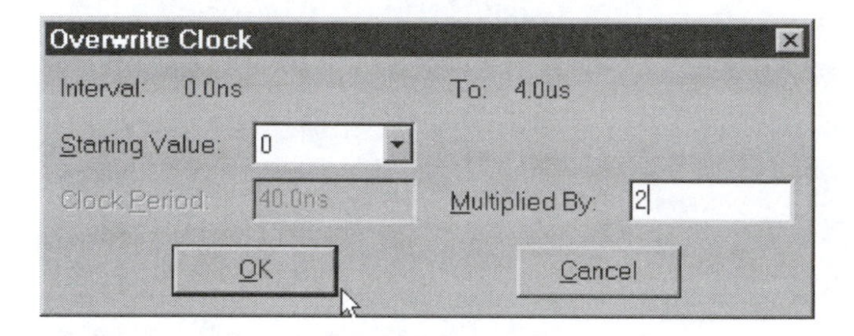

Figure 7.4 Applying a Clock Waveform

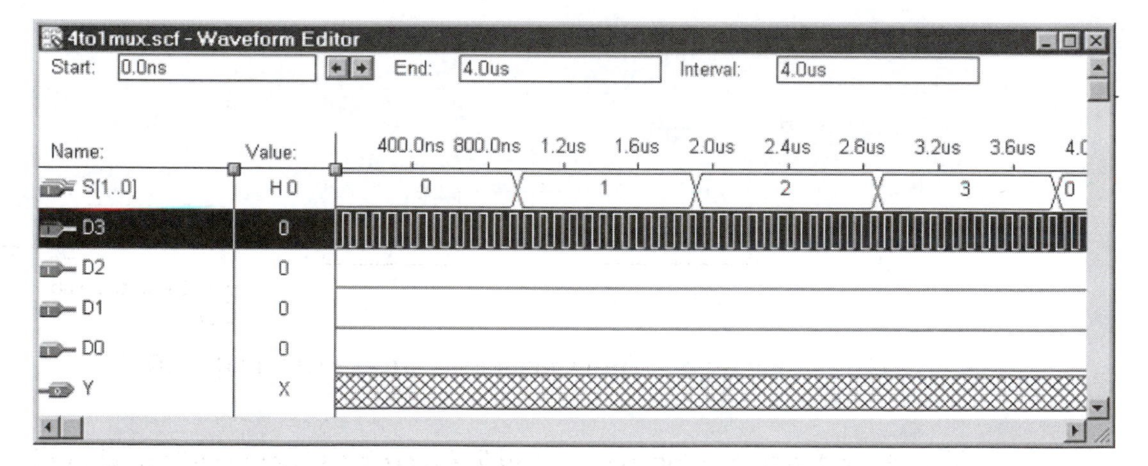

Figure 7.5 Effect of Adding a Clock Waveform

Repeat the previous step for D_2 (4 times base), D_1 (1 times base), and D_0 (8 times base). These frequencies were chosen to make as great a contrast as possible between adjacent inputs so that the different selected inputs could easily be seen.

3. Run the simulation and show the waveforms to your instructor.

Instructor's Initials: ________

Testing the 4-to-1 MUX

One way to test the MUX function on the Altera UP-1 board is to apply a known signal to each data input, as we did in our simulation, and manually change the values of the select inputs with DIP switches. The output signal can be observed on a monitoring device such as an LED. The LED behavior will tell you which MUX input channel has been selected.

Two MUX test circuits are shown in Figures 7.6 and 7.7. The test circuits are examples of **hierarchical design.** All this means is that the test circuit (called the **top level** of the hierarchy) contains components that are complete designs in and of themselves. The design hierarchy shown contains two components: a 4-to-1 multiplexer, created from the graphic design file in Figure 7.1, and a predesigned clock divider, **clkdiv1** or **clkdiv2**.

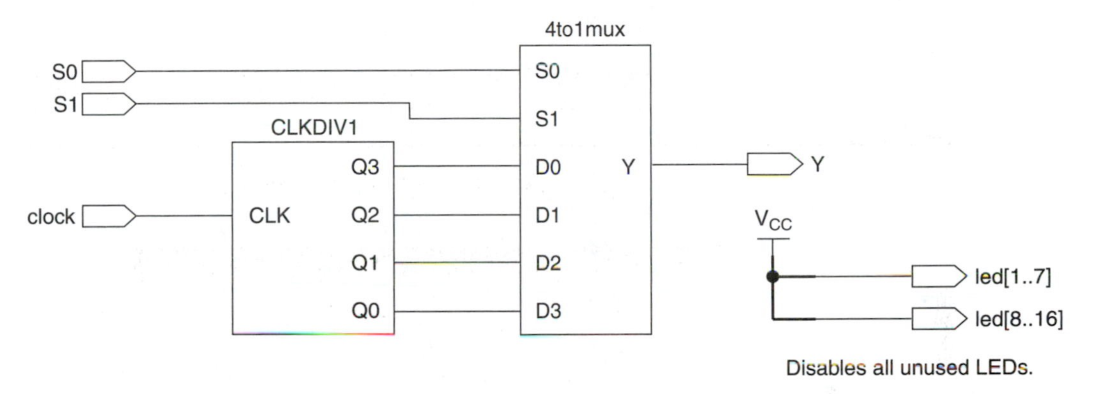

Figure 7.6 Test Circuit for a 4-to-1 MUX (Altera UP-1)

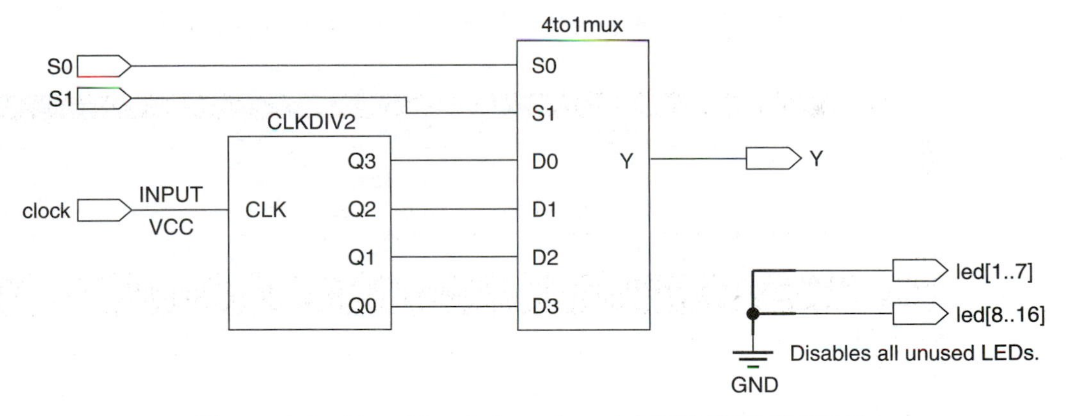

Figure 7.7 Test Circuit for a 4-to-1 MUX (RSR PLDT-2)

The clock dividers in Figures 7.6 and 7.7 (**clkdiv1** (Altera UP-1) or **clkdiv2** (RSR PLDT-2)) provide digital square wave signals of binary-multiple frequencies at four different outputs (Altera UP-1: 1.5 Hz, 3 Hz, 6Hz, and 12 Hz; RSR PLDT-2: 1 Hz, 2 Hz, 4 Hz, 8 Hz). These frequencies are slow enough to be observed visually.

To create the test circuit, we must first create the symbols for the components, as follows.

1. Open the graphic design file **4to1mux.gdf** and set the project to the current file. From the file menu, select **Create Default Symbol** (Figure 7.8). A symbol file may already exist, as it is normally created by the compile process. If so, the dialog box in Figure 7.9 will appear. Click **OK**.

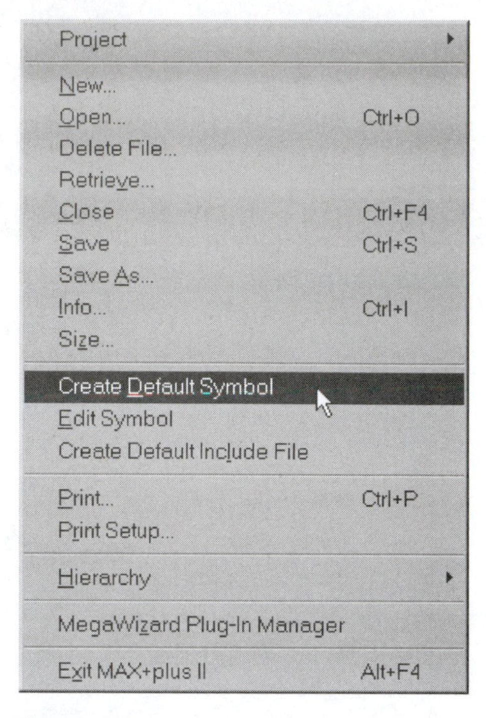

Figure 7.8 Create Default Symbol (File Menu)

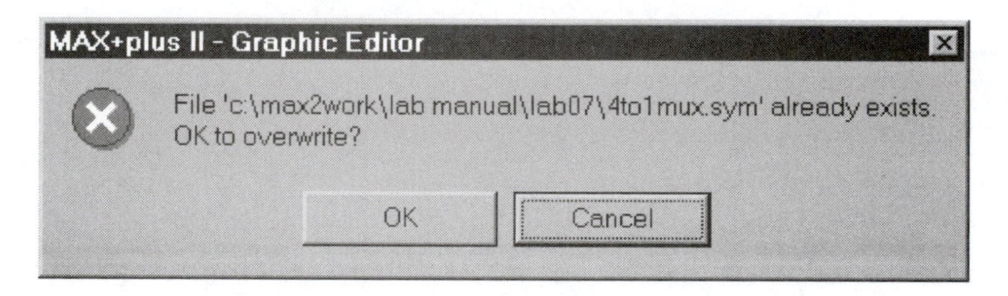

Figure 7.9 Overwriting a Symbol File

2. Copy the files from the **lab07** folder in the student files on the accompanying CD to the folder ***drive:*\max2work\lab07**. Open the file **clkdiv1.vhd** (Altera UP-1) or **clkdiv2.vhd** (RSR PLDT-2). Set the project to the current file and create a default symbol for the clock divider.

3. Create a new graphic design file and save it as ***drive:*\max2work\lab07\test4to1_1.gdf** (Altera UP-1) or ***drive:*\max2work\lab07\test4to1_2.gdf** (RSR PLDT-2). Set the project to the current file.

When saving the top-level file of a MAX+PLUS II design hierarchy, make sure that you save it under its own unique name. **Do not just click OK when saving the file for the first time** because you may inadvertently give the top-level file the same name as an existing component. **Do not UNDER ANY CIRCUMSTANCES** give the top-level file in a hierarchy the same name as a component in that hierarchy. (e.g., Do not name the top level file **4to1mux** if it contains a component called **4to1mux**.) This creates a condition known as a recursive hierarchy tree, which is the graphical equivalent of an infinite loop. This situation is very difficult to fix, once created. To fix it, you may need to delete all project files, except for the top-level and component design files (gdf and vhd), rename the design file for the offending component, create a new component symbol, replace the component in the top-level file, and recompile the design.

4. Enter the component symbols from the dialog box shown in Figure 7.10. Connect the components and assign the pin names as shown in Figure 7.6 or 7.7.

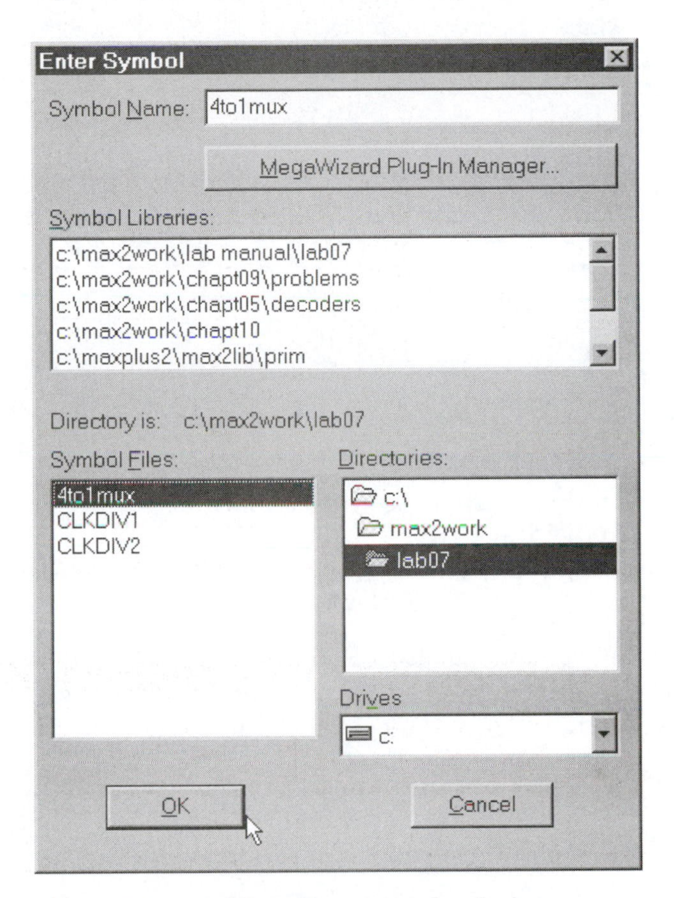

Figure 7.10 Adding Compnent Symbols to a Graphic Design File

Create two outputs for disabling unused LEDs: **led[1..7]** and **led[9..16]**. For the Altera UP-1 board, connect these LED outputs to V_{CC}. For the RSR PLDT-2 board, connect the LED outputs to ground. Draw the connecting line from the output pin symbol to the VCC or GND symbol, not the other way around. Make sure that the connecting line is a thick line, indicating a connection to multiple output lines for one pin symbol. The line thickness can also be changed by highlighting the line, then right-clicking it and choosing Line Style from the pop-up menu. Save and compile the file.

5. Once the design hierarchy has been created, the component files can be accessed for editing by double-clicking. Try double-clicking on the component called **4to1mux**.

What do you see? ____

____

6. Bring the design file for the test circuit to the front window and set the project to the current file. Assign a device (EPM7128SLC84). Assign pin numbers as shown in Table 7.3. When you have assigned the pin numbers, save and compile the file again.

Table 7.3 Pin Assignments for 4-to-1 MUX Test Circuit

Function	Pin Number
s1 (SW1-7)	39
s0 (SW1-8)	41
clock	83
y (LED8)	52
LED1	44
LED2	45
LED3	46
LED4	48
LED5	49
LED6	50

Function	Pin Number
LED7	51
LED9	80
LED10	81
LED11	4
LED12	5
LED13	6
LED14	8
LED15	9
LED16	10

7. Connect short lengths of #22 solid-core wire from the prototyping headers around the EPM7128S chip to two DIP switches (for S_1 and S_0) and the LEDs, as required.

   On the RSR board, if no wire connections are in place, it is not necessary to do any wiring. On the Altera UP-1 board, if the wire connections are not in place, it is only necessary to connect S_1, S_0, and Y. The CLOCK connection is hardwired on both boards.

   The pin assignments for LED1 through LED7 and LED9 through LED16 are there only so that these LEDs are not lit if they have been wired for another project, so don't connect them if they are not already connected.

8. Download the test circuit to the CPLD test board. Set the S_1S_0 switches to 00 and observe the output of the MUX on the output LED. Repeat for values of 01, 10, and 11. Explain your observations to your instructor.

Instructor's Initials: __________

VHDL Multiplexer

1. Write a VHDL file, using a selected signal assignment statement, to design an 8-to-1 multiplexer. Define the data inputs and select inputs as BIT_VECTOR types. Save the file as *drive:***max2work\lab07\mux_8ch.vhd.** For the Altera UP-1 board, make the MUX output active-LOW so that it can drive an active-LOW LED. For the RSR PLDT-2 board, make the output active-HIGH.

2. Create a simulation file for the 8-to-1 mux to verify the design operation. Show the waveforms to your instructor.

Instructor's Initials: __________

3. Make a new Graphic Design File called *drive:***max2work\lab07\mux8test_1.gdf** (Altera UP-1) or *drive:***max2work\lab07\mux8test_2.gdf** (RSR PLDT-2), as shown in Figures 7.11 and 7.12.

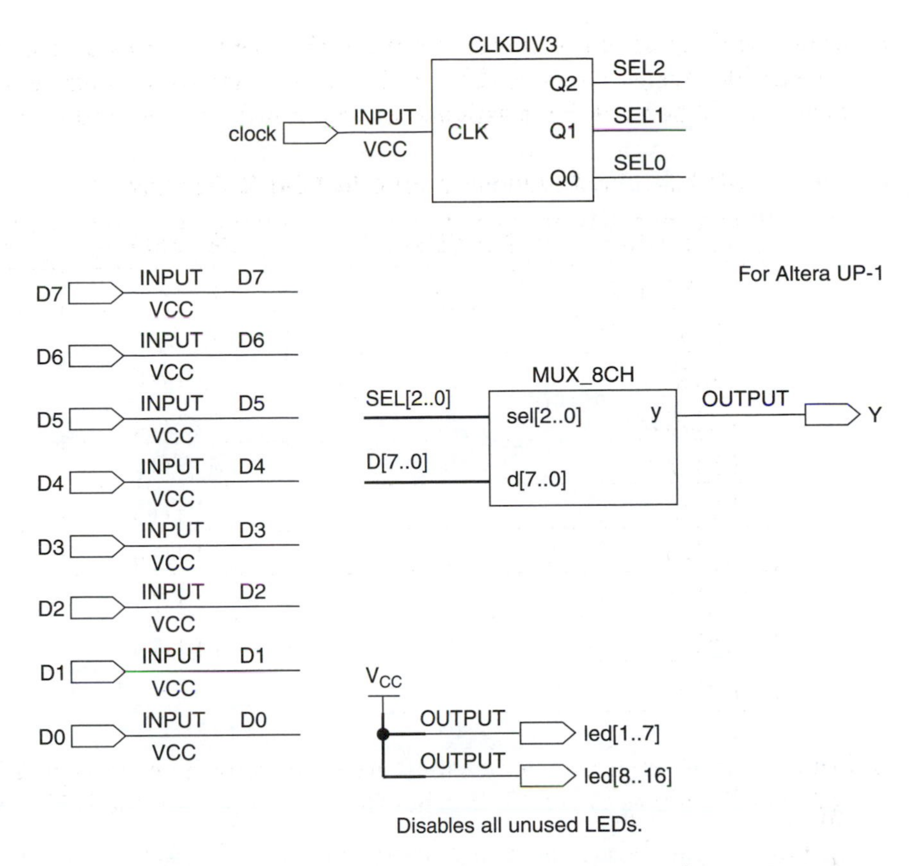

Figure 7.11 Test Circuit for an 8-to-1 MUX (Altera UP-1)

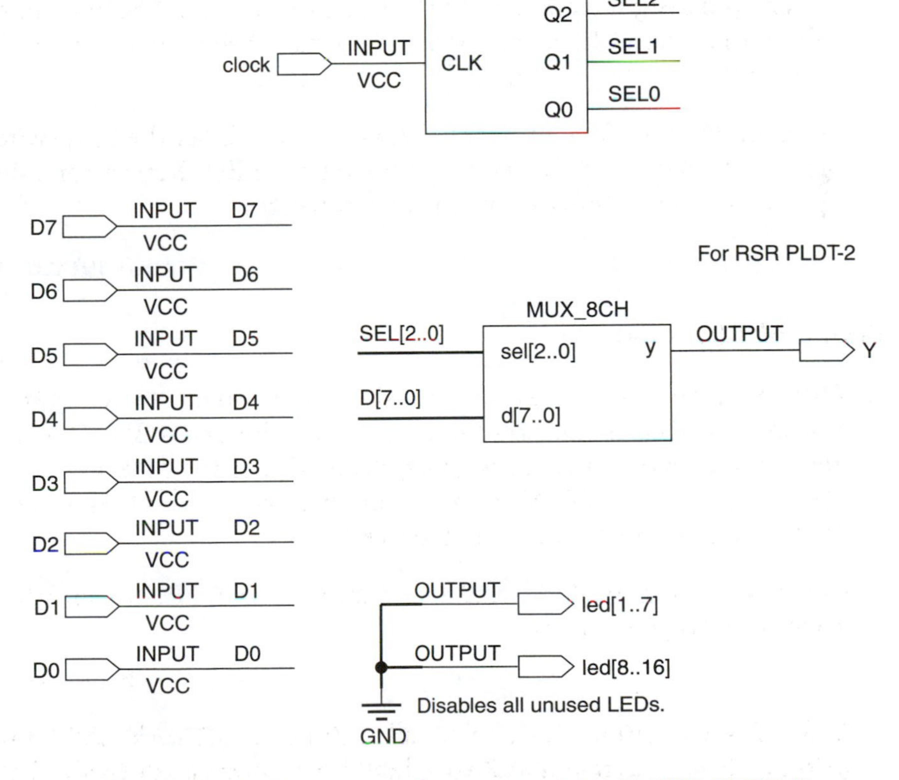

Figure 7.12 Test Circuit for an 8-to-1 MUX (RSR PLDT-2)

Note that you can connect two nodes (i.e., two points in the circuit) by labeling each node with the same name. If you want to connect individual nodes to a bus, label each node with a name and number (D0, D1, D2,..,D7) and label the bus with a group name, with the range of input labels in square brackets (D[7..0]). To enter the names, highlight a node or bus and right-click the mouse. Select **Enter Node/Bus Name** from the pop-up menu that appears and enter the label text.

4. Assign a device (EPM7128SLC84) for the design. Assign pins as shown in Table 7.4.

Table 7.4 Pin Assignment for an 8-to1 MUX Test Circuit (Pattern Generator)

Function	Pin Number
D7 (SW1-1)	34
D6 (SW1-2)	33
D5 (SW1-3)	36
D4 (SW1-4)	35
D3 (SW1-5)	37
D2 (SW1-6)	40
D1 (SW1-7)	39
D0 (SW1-8)	41
clock	83
Y (LED8)	52
LED1	44
LED2	45
LED3	46

Function	Pin Number
LED4	48
LED5	49
LED6	50
LED7	51
LED9	80
LED10	81
LED11	4
LED12	5
LED13	6
LED14	8
LED15	9
LED16	10

5. If using the Altera UP-1 board, connect the D inputs on the 8-to-1 MUX to the DIP switches SW1-1 to SW1-8, using short lengths of #22 solid-core wire. Connect the Y output to LED8. The connections are already made on the RSR PLDT-2 board by the installed jumpers. Download the design to the CPLD on the Altera UP-1 board or RSR PLDT-2 board.

6. The circuit in Figures 7.11 and 7.12 can be used as a pattern generator; that is, a circuit to generate a repetitive flashing pattern on the output LED. The clock divider component (**clkdiv3** (Altera UP-1) or **clkdiv4** (RSR PLDT-2)) creates a repeating binary sequence from 000 to 111 on the MUX select inputs. The result is that the pattern applied to the MUX data inputs (as supplied from the DIP switch inputs) will be applied in sequence to the output LED.

   Set the DIP switches to an alternating pattern of 0s and 1s: 01010101. What do you observe?

   ___

   ___

   ___

Change the pattern to 01000101. Note your observations.

____

____

____

____

Try other combinations such as 00110011, 00001111, 11111110 and 11111010. What do you see?

____

____

____

____

____

____

Demonstrate the operation of the circuit to your instructor.

Instructor's Initials: ________

VHDL Bus Multiplexer

1. Write a VHDL file that defines a multiplexer that switches two 4-bit inputs, **x** and **y** to a 4-bit output, **z**. Define **x**, **y**, and **z** as type BIT_VECTOR. Save the file as *drive:*\max2work\lab07\quad2to1.vhd.
2. Create a simulation for the Quadruple 2-to-1 MUX. Show the simulation to your instructor.

   Instructor's Initials: ________

3. Create a default symbol for the MUX and, in a new **gdf**, connect the **z** outputs to a hexadecimal-to-seven-segment decoder created in Lab 6, as shown in Figures 7.13 (page 70) and 7.14 (page 71). Recall that the Altera UP-1 board has common anode (active-LOW) numerical displays and the RSR PLDT-2 board has common cathode (active-HIGH) displays.
4. Assign the device as EPM7128SLC84. Assign pin numbers as shown in Table 7.5.

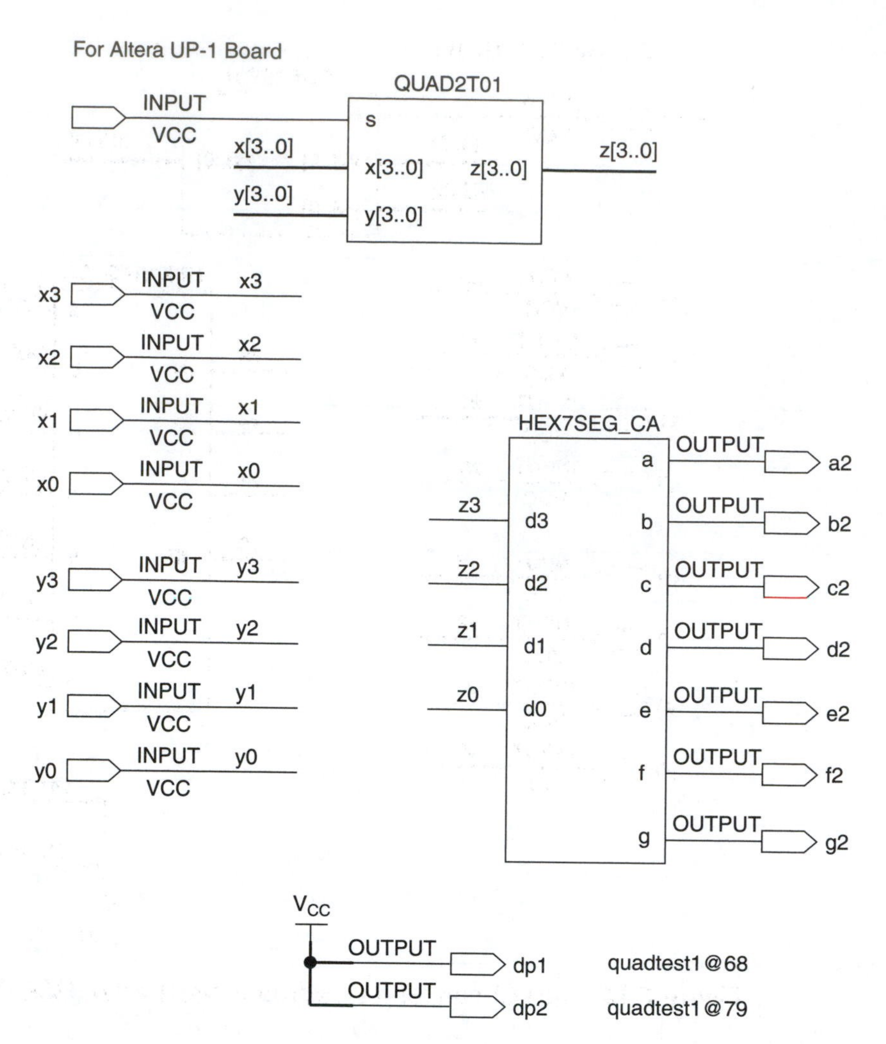

Figure 7.13 Test Circuit for a Quadruple 2-to-1 MUX (Altera UP-1)

Table 7.5 Pin Assignments for a Quadruple 2-to-1 MUX and Display

Function	Pin Number
x3 (SW1-1)	34
x2 (SW1-2)	33
x1 (SW1-3)	36
x0 (SW1-4)	35
y3 (SW1-5)	37
y2 (SW1-6)	40
y1 (SW1-7)	39
y0 (SW1-8)	41
Select (PB1)	11

Function	Pin Number
a2	69
b2	70
c2	73
d2	74
e2	76
f2	75
g2	77
dp2	79
dp1	68

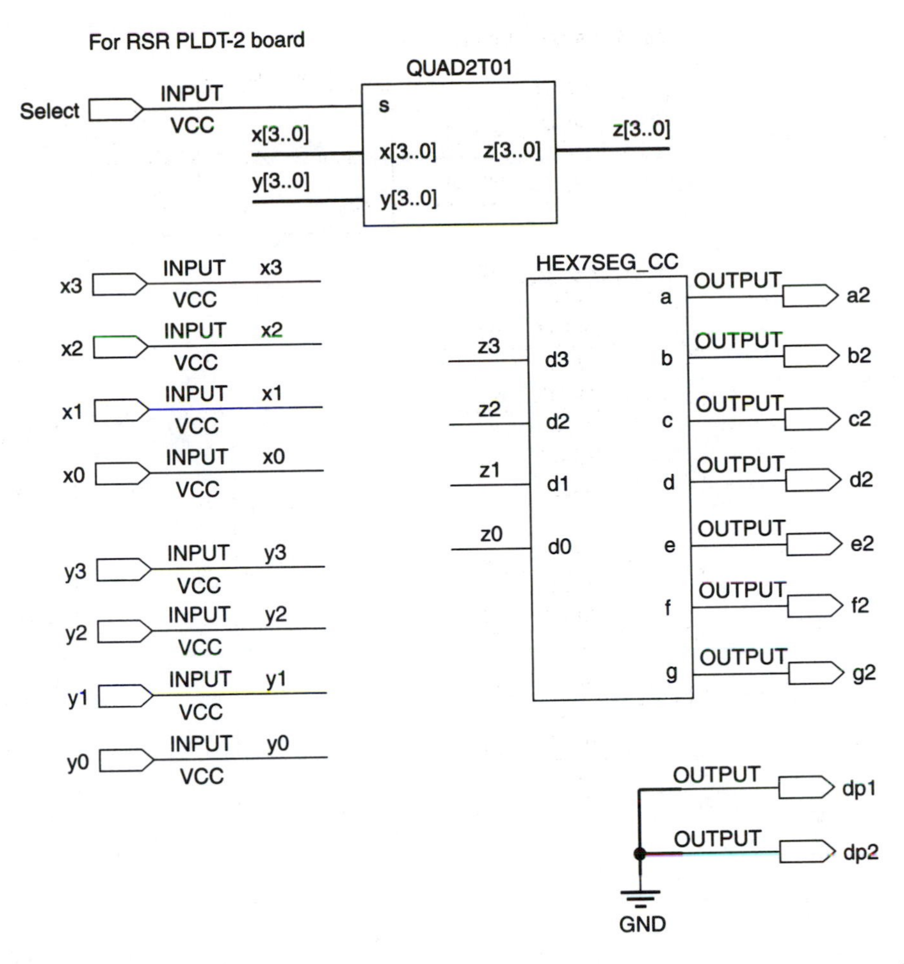

Figure 7.14 Test Circuit for a Quadruple 2-to-1 MUX (RSR PLDT-2 Board)

5. Save and compile the file and download it to the Altera UP-1 or RSR PLDT-2 board. Connect the pin jacks for the DIP switches to the prototyping headers for the EPM7128S chip on the Altera UP-1 board. The connections on the RSR PLDT-2 board are made by the installed jumpers. Connect pushbutton PB1 to pin 11 on the CPLD header.

6. Set switches SW1-1 to SW1-4 (input **x**) to 0101. Set switches SW1-5 to SW1-8 (input **y**) to 1110. What effect does the **Select** switch (PB1) have on the digit shown on the seven-segment display? Show the results to your instructor.

Instructor's Initials: ___________

Assignment Questions

Complete problems 5.26 and 5.27 in *Digital Design with CPLD Applications and VHDL*.

Lab 8

Combinational Logic Functions: Priority Encoder and Magnitude Comparator

Name ______________________ Class __________ Date ______________

Objectives Upon completion of this laboratory exercise, you should be able to:

- Enter a VHDL design file and create a simulation for a BCD priority encoder.
- Enter a test circuit for a BCD priority encoder in the MAX+PLUS II graphic editor.
- Create a 3-bit magnitude comparator as a MAX+PLUS II Graphic Design File and as a VHDL design entity.
- Enter test circuits for the magnitude comparators in the MAX+PLUS II graphic editor.

Reference Dueck, Robert K., *Digital Design with CPLD Applications and VHDL*

Chapter 5: Combinational Logic Functions
5.2 Encoders
5.5 Magnitude Comparators

Equipment Required CPLD Trainer:

Altera UP-1 Circuit Board with ByteBlaster Download Cable, or
RSR PLDT-2 Circuit Board with Straight-Through Parallel Port Cable, or
Equivalent CPLD Trainer Board with Altera EPM7128S CPLD

MAX+PLUS II Student Edition Software
AC Adapter, minimum output: 7 VDC, 250 mA DC
Anti-static wrist strap
#22 solid-core wire
Wire strippers

Experimental Notes

In this lab, we will look at two combinational logic circuits: a **priority encoder** and a **magnitude comparator**. Both circuits can be easily implemented in VHDL and with more difficulty in a Graphic Design File. Both circuits become very much more complex with each additional bit, so it is often easiest to use various VHDL constructs that describe the behavior of each circuit, rather than its specific Boolean equations.

Priority Encoder

A priority encoder is a circuit that generates a binary or BCD code that corresponds to the active input with the highest priority, which generally means the input with the highest decimal subscript. Figure 8.1 shows an example of the operation of a BCD priority encoder.

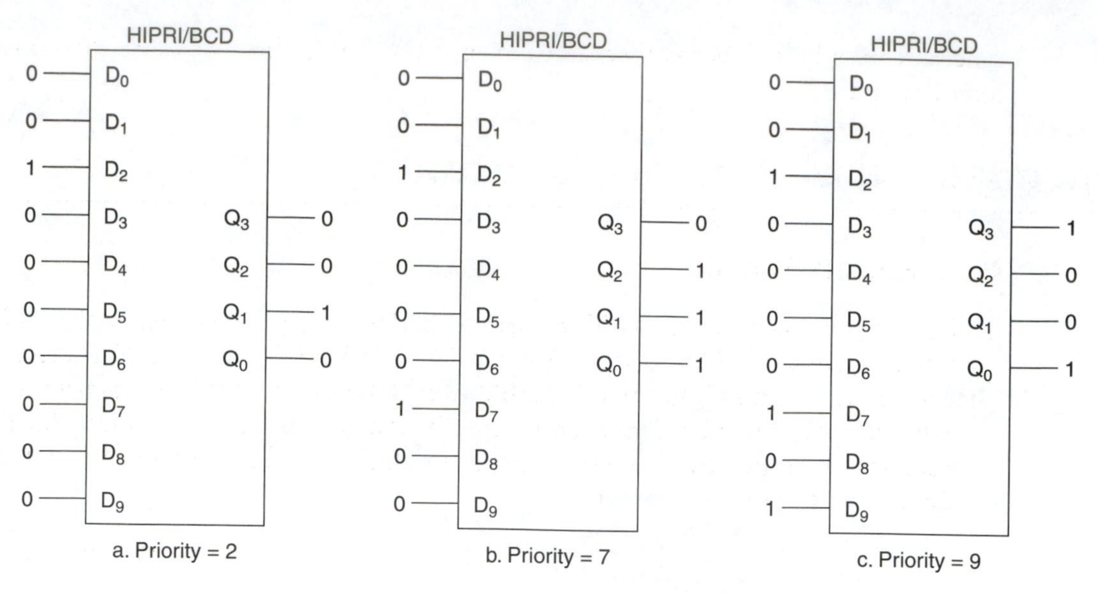

Figure 8.1 BCD Priority Encoder

In Figure 8.1a, only input D_2 is active, so the output code generated is 0010, the binary equivalent of 2. In Figure 8.1b, input D_2 remains active, but now D_7 is also active. Since D_7 has the higher priority, the output code is now 0111, the binary equivalent of 7. In Figure 8.1c, input D_9 is HIGH, in addition to the other two inputs. Since D_9 has the highest priority, the output is 1001.

Magnitude Comparator

A magnitude comparator accepts two binary numbers of equal width and generates outputs that indicate whether the numbers are equal, and if not, which is greater. Figure 8.2 shows the operation of a magnitude comparator that compares two 3-bit numbers, A and B, and determines whether $A = B$, $A > B$, or $A < B$. A true value for any of these conditions makes the appropriate output HIGH.

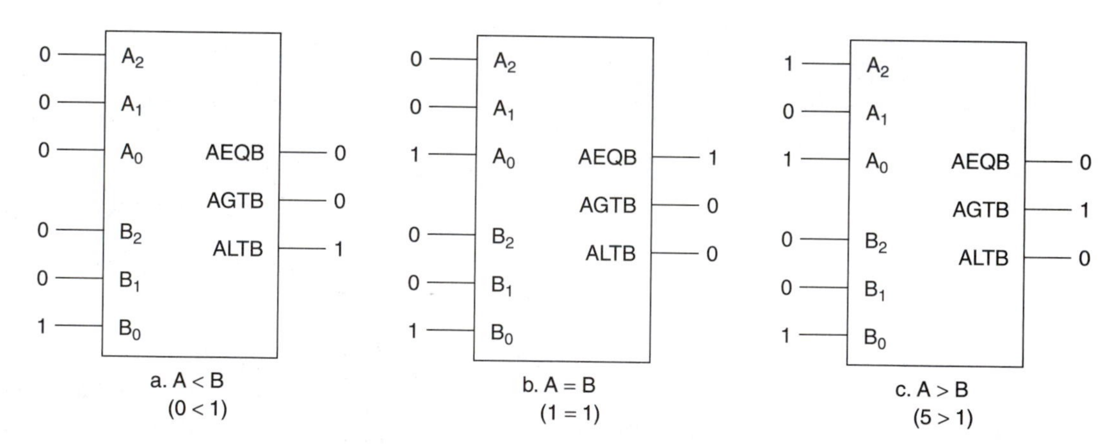

Figure 8.2 3-bit Magnitude Comparator

Figure 8.2a shows $A_2A_1A_0 = 000$ and $B_2B_1B_0 = 001$. Since $0 < 1$, the ALTB (A Less Than B) output goes HIGH and the other two outputs remain LOW. Figure 8.2b indicates that $A_2A_1A_0 = 001$ and $B_2B_1B_0 = 001$. Since $1 = 1$, the AEQB (A EQuals B) output goes HIGH and the other two outputs are LOW. In Figure 8.2c, $A_2A_1A_0 = 101$ and $B_2B_1B_0 = 001$. Since $5 > 1$, the AGTB (A Greater Than B) output goes HIGH and the other two outputs are LOW. Logically, only one output can be HIGH at any time, since *A* can only be less than, equal to, or greater than *B*, but not more than one of these simultaneously.

Procedure

BCD Priority Encoder

1. Refer to the section on VHDL priority encoders in *Digital Design with CPLD Applications and VHDL* (pp. 182–185). Write a VHDL file for a BCD priority encoder and save it as ***drive:*\max2work\lab08\hipri_10.vhd**. Do not use concurrent signal assignment statements (i.e., do not directly code the Boolean equations for the encoder). Compile the file and create a simulation for the encoder. Show the simulation to your instructor.

Instructor's Initials: __________

2. Figures 8.3 and 8.4 show test circuits for the BCD priority encoder for the Altera UP-1 board and the RSR PLDT-2 board, respectively. The test circuit displays the output of the priority encoder both as a binary value and as a decimal digit on the board's seven-segment numerical display. Unused LEDs are disabled by tying them either to V_{CC} or ground.

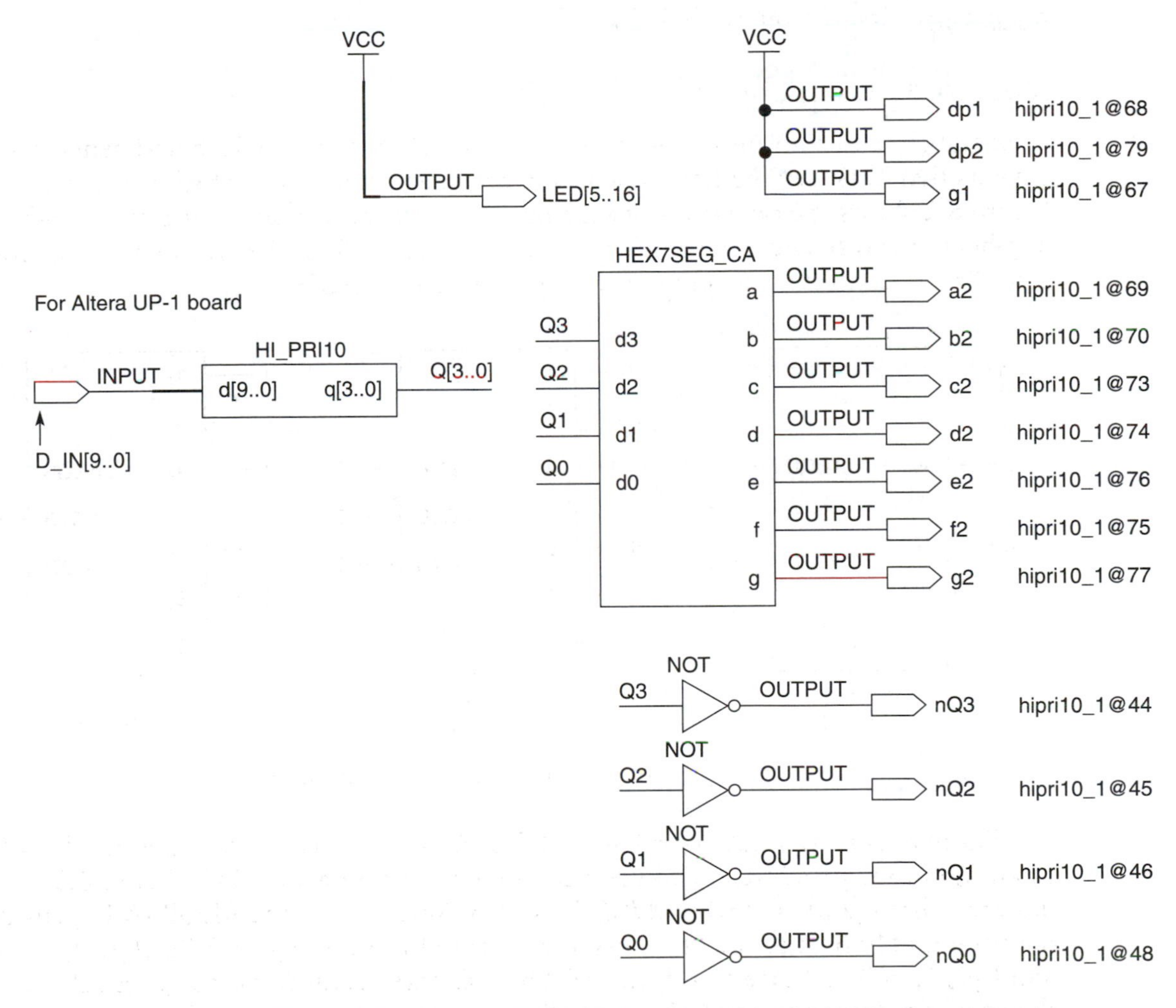

Figure 8.3 Priority Encoder Test Circuit (Altera UP-1)

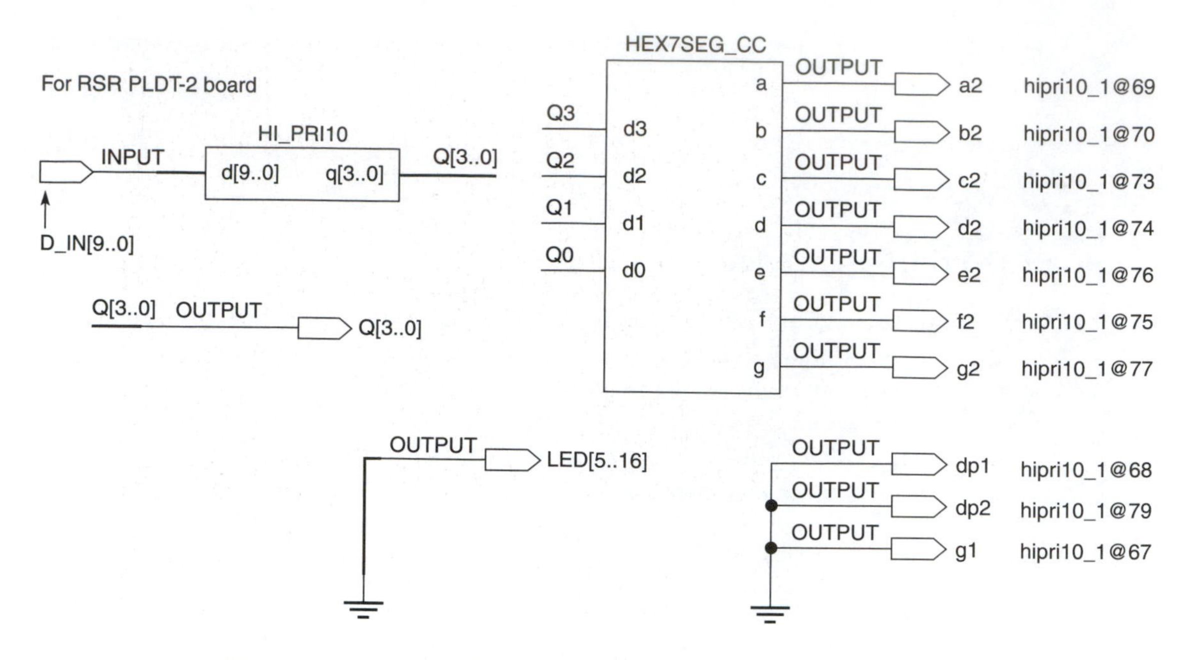

Figure 8.4 Priority Encoder Test Circuit (RSR PLDT-2 Board)

Use the seven-segment decoder from lab 6 in the test circuit (common anode for the Altera UP-1, common cathode for the RSR PLDT-2). You do not have to copy the seven-segment decoder file (**hex7seg_ca.vhd** or **hex7seg_cc.vhd**) to the **lab08** folder if you assign a user library to point to the folder holding the VHDL design file. (A user library is simply a file path telling MAX+PLUS II where to look for a design file. The default path is the current directory, followed by listed user libraries in prioritized order. Refer to pp. 130–132 in *Digital Design with CPLD Applications and VHDL* for further details.)

To assign a user library, select **User Libraries** from the **Options** menu, shown in Figure 8.5. The dialog box in Figure 8.6 appears. Select the ***drive:*\max2work\lab06** folder by double-clicking in the box labeled **Directories.** Click **Add** to transfer this folder name to the **Existing Directories** box. Click **OK**. From now on, any design file from the folder ***drive:*\ max2work\lab08** can be used in a MAX+PLUS II design hierarchy without copying it to the same folder as the top-level design file. (If this does not work, copy the vhd file for the seven-segment decoder to the **maxzwork\lab08** folder.)

Assign the device (EPM7128SLC84) and pin numbers as shown. Assign pin numbers to the unused LEDs according to Table 8.1 at the end of this lab. The Q outputs for the RSR PLDT-2 board are the same as for the nQ outputs on the Altera UP-1 board.

Compile the file and download it to your CPLD board. Demonstrate the circuit operation to your instructor.

Instructor's Initials: __________

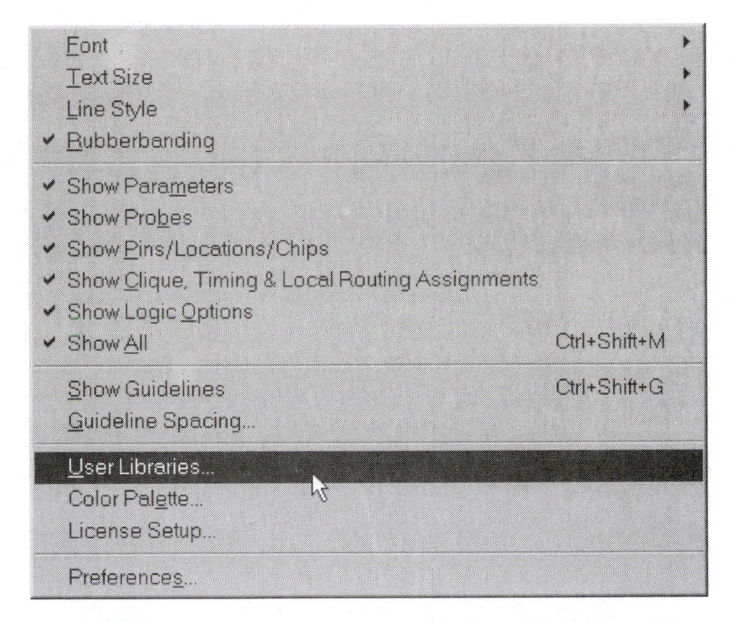

Figure 8.5 Options Menu (User Libraries)

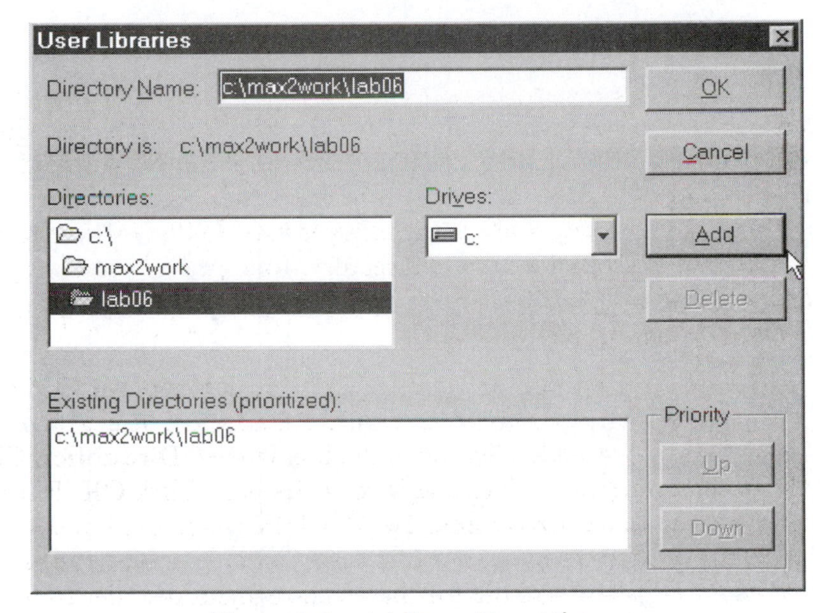

Figure 8.6 Adding a User Library

Magnitude Comparator (Graphic Design File)

1. Figure 8.7 on page 78 shows the logic diagram of a circuit that implements the 3-bit magnitude comparator of Figure 8.2. Write the Boolean equations for this circuit.

   AEQB = ______________________________

   AGTB = ______________________________

   ALTB = ______________________________

2. Use the MAX+PLUS II graphic editor to create the circuit of Figure 8.7. Save the file as ***drive:*\max2work\lab08\3bit_cmp.gdf**. Set the project to the current file. Assign the device as EPM7128SLC84-7 or EPM7128SLC84-15, depending on which device is

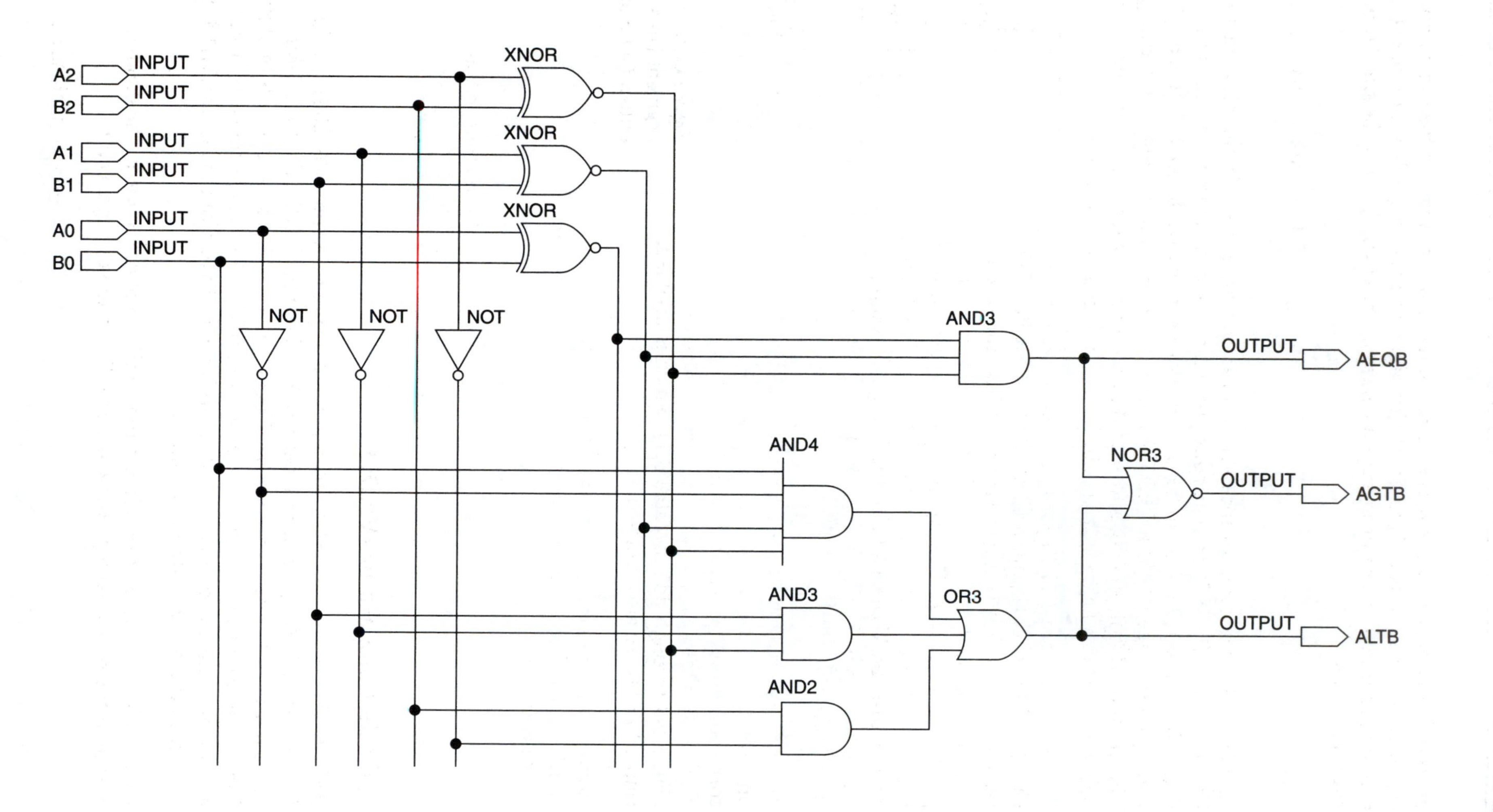

Figure 8.7 3-bit Magnitude Comparator

in your CPLD board. Compile the design and create a simulation file that verifies the correctness of the design. (The simulation file may contain glitches, or noise spikes, that are generated due to propagation delays in the circuit.) Show the simulation to your instructor.

Instructor's Initials: ________

3. Create a default symbol for the 3-bit comparator you entered in the previous step. Use the MAX+PLUS II graphic editor to create the comparator test circuit in Figure 8.8 on page 80 (Altera UP-1 board: ***drive:*\max2work\lab08\3bit_cmp_1.gdf**) or Figure 8.9 on page 81 (RSR PLDT-2 board: ***drive:*\max2work\lab08\3bit_cmp_2.gdf**). The outputs **AEQB**, **AGTB**, and **ALTB** turn on an LED for a true condition. Note that only one LED can be on at a time if the circuit is functioning correctly.

   The components **hi_lo_ca** (common anode for Altera UP-1) and **hi_lo_cc** (common cathode for RSR PLDT-2) are special decoders that generate an "L" on the seven-segment display if $A < B$, an "H" if $A > B$, and three parallel bars if $A = B$. The segment patterns are shown in Figure 8.10. Any other condition (if it existed) will cause a blank display. Write a VHDL file for this function, create a component (**hi_lo_ca** or **hi_lo_cc**), and add it to the test circuit.

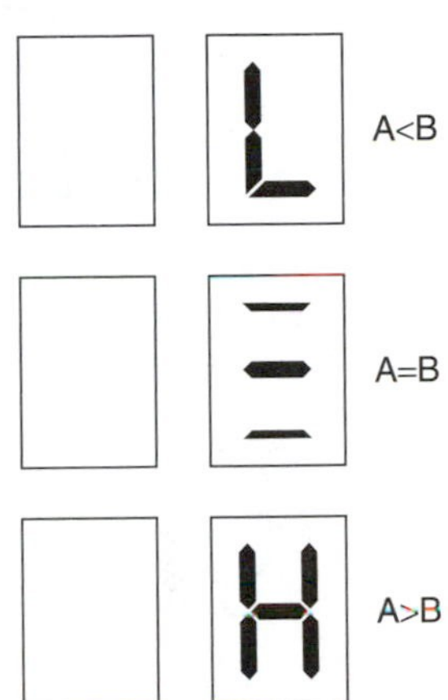

Figure 8.10 Seven-Segment Display output Generated by Hi_Lo_CC and Hi_Lo_CA

4. Assign pin numbers to the design, according to Table 8.1 on page 84 at the end of this lab exercise. **Outputs[1..8]** correspond to seven-segment digit 2 (**a2** through **dp2** in Table 8.1). **Outputs[9..16]** correspond to seven-segment digit 1 (**a1** through **dp1** in Table 8.1), which is disabled. Unused LEDs are also disabled.

5. Compile the test circuit design and download it to your CPLD board. Demonstrate the operation of the circuit to your instructor.

Instructor's Initials: ________

Magnitude Comparator (VHDL)

1. Refer to the section on VHDL Magnitude Comparators (pp. 206–208) in *Digital Design with CPLD Applications and VHDL*. Write the VHDL file for a 3-bit comparator with outputs for $A = B$ (AEQB), $A > B$ (AGTB), and $A < B$ (ALTB). Design the comparator outputs to be active-HIGH. Do not use concurrent signal assignment statements. Save the file as ***drive:*\max2work\lab08\compare3.vhd**. Set the project to the current file and compile.

2. Create a MAX+PLUS II simulation that verifies the operation of the comparator design in the previous step. Show the simulation to your instructor.

Instructor's Initials: ________

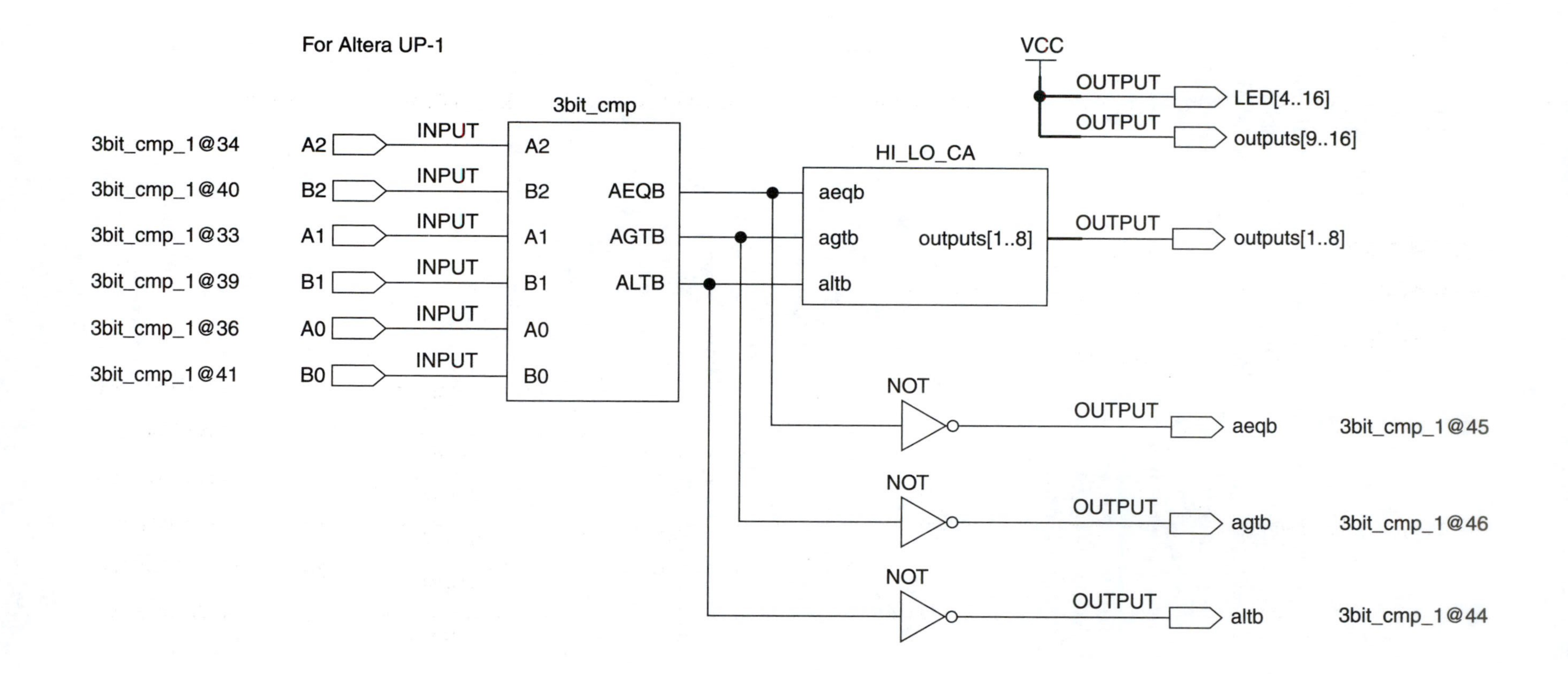

Figure 8.8 3-Bit Comparator Test Circuit (Graphics Design File, Altera UP-1)

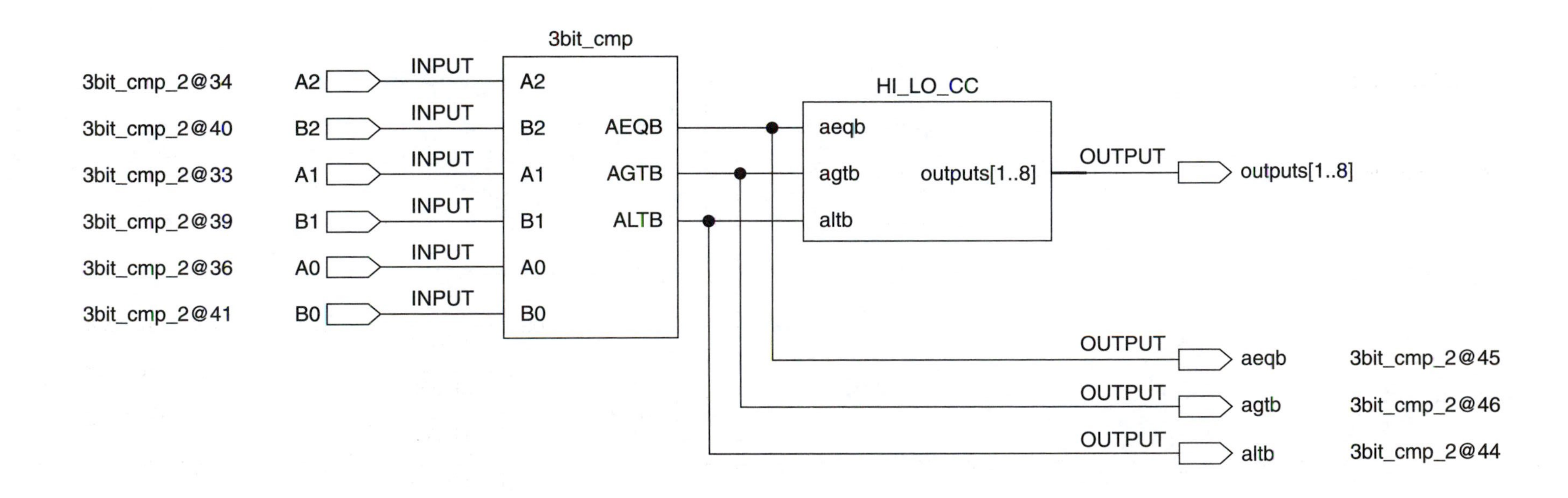

FIGURE 8.9 3-bit Comparator Test Circuit (Graphics Design File, RSR PLDT-2)

3. Create a default symbol for the VHDL comparator. Use the MAX+PLUS II graphic editor to create a test circuit for the VHDL 3-bit comparator.

   A test circuit for the Altera UP-1 board is shown in Figure 8.11. Save this file as *drive:*\max2work\lab08\3bit_cmp_3.gdf. Figure 8.12 shows a test circuit for the RSR PLDT-2 board. Save this as *drive:*\max2work\lab08\3bit_cmp_4.gdf. The pin numbers are identical to the test circuit based on the gdf component in the previous section.

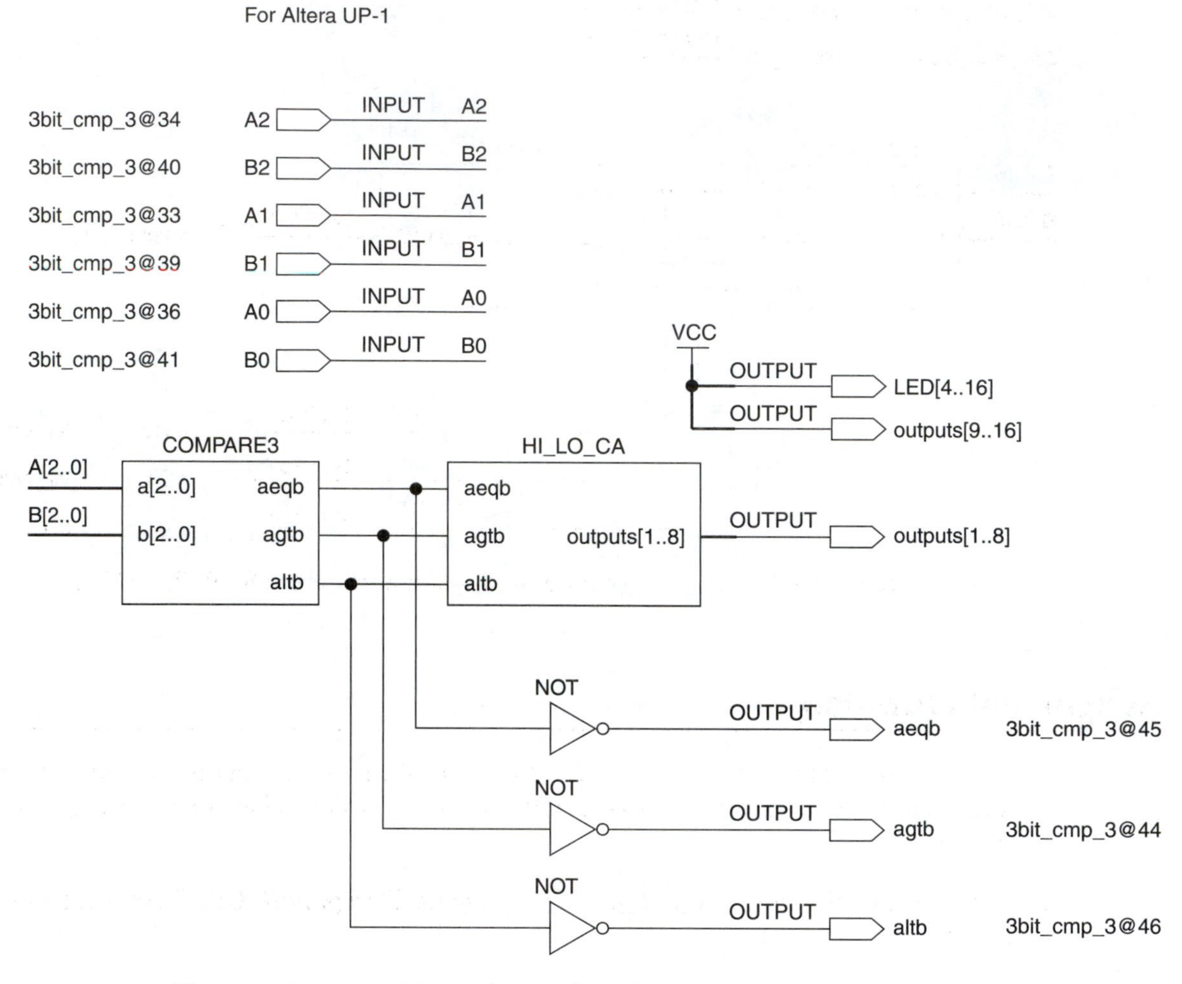

Figure 8.11 Test Circuit for a 3-bit Comparator (VHDL, Altera UP-1)

Tip You can save the previous test circuit under the new name and modify it to include the VHDL comparator, rather than gdf comparator. (This will save you the trouble of entering the pin numbers all over again.)

Demonstrate the circuit operation to your instructor.

Instructor's Initials: __________

For RSR PLDT-2

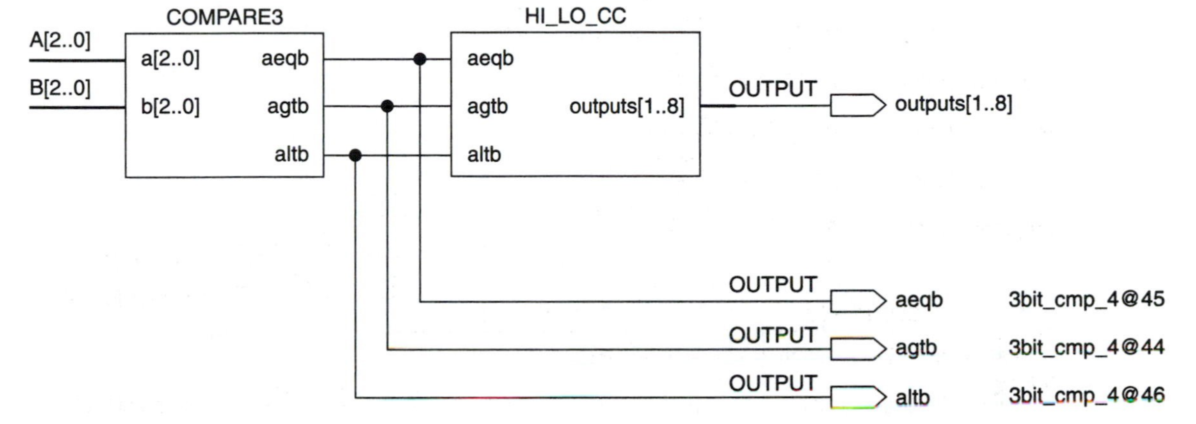

Figure 8.12 Test Circuit for a 3-bit Comparator (VHDL, RSR PLDT-2)

Assignment Questions

1. Briefly explain the operation of each function of the 3-bit magnitude comparator in Figure 8.7. Refer to the Boolean equations of the circuit when formulating your answer.

2. Complete problems 5.18 and 5.44 from *Digital Design with CPLD Applications and VHDL.*

Table 8.1 EPM7128LC84 Pin Assignments
Altera UP-1 Board and PLDT-2 Board

Seven Segment Digits			
Function	**Pin**	**Function**	**Pin**
a1	58	a2	69
b1	60	b2	70
c1	61	c2	73
d1	63	d2	74
e1	64	e2	76
f1	65	f2	75
g1	67	g2	77
dp1	68	dp2	79

Pushbuttons			
Function	**Pin**	**Function**	**Pin**
PB1	11	PB2	1

DIP Switches			
Function	**Pin**	**Function**	**Pin**
SW1-1	34	SW2-1	28
SW1-2	33	SW2-2	29
SW1-3	36	SW2-3	30
SW1-4	35	SW2-4	31
SW1-5	37	SW2-5	57
SW1-6	40	SW2-6	55
SW1-7	39	SW2-7	56
SW1-8	41	SW2-8	54

LED Outputs			
Function	**Pin**	**Function**	**Pin**
LED1	44	LED9	80
LED2	45	LED10	81
LED3	46	LED11	4
LED4	48	LED12	5
LED5	49	LED13	6
LED6	50	LED14	8
LED7	51	LED15	9
LED8	52	LED16	10

Unassigned: Pins 12, 15, 16, 17, 18, 20, 21, 22, 24, 25, 27

Special Function: Pin 1 (GCLRn); Pin 2 (Input/OE2/GCLK2); Pin 83 (GCLK1, hardwired); Pin 84 (OE1)

Lab 9

Arithmetic Circuits in VHDL

Name ______________________ Class __________ Date ______________

Objectives Upon completion of this laboratory exercise, you should be able to:

- Create and simulate a full adder in VHDL, assign pins to the design, and test it on a CPLD circuit board.
- Use a VHDL full adder as a component in an 8-bit two's complement adder/subtractor.
- Create a VHDL hierarchical design, including components for full adders and seven-segment decoders, without using the MAX+PLUS II Graphic Editor.
- Design an overflow detector for use in a VHDL two's complement adder/subtractor.

Reference Dueck, Robert K., *Digital Design with CPLD Applications and VHDL*

Chapter 6: Digital Arithmetic and Arithmetic Circuits
6.1 Digital Arithmetic
6.2 Representing Signed Binary Numbers
6.3 Signed Binary Arithmetic
6.6 Binary Adders and Subtractors

Equipment Required CPLD Trainer:
Altera UP-1 Circuit Board with ByteBlaster Download Cable, or
RSR PLDT-2 Circuit Board with Straight-Through Parallel Port Cable, or
Equivalent CPLD Trainer Board with Altera EPM7128S CPLD
MAX+PLUS II Student Edition Software
AC Adapter, minimum output: 7 VDC, 250 mA DC
Anti-static wrist strap
#22 solid-core wire
Wire strippers

Experimental Notes

Note The designs in this laboratory exercise are to be done entirely in VHDL, without using the MAX+PLUS II Graphic Editor for any portion of the design.

Arithmetic Circuits

Circuits for performing binary arithmetic are based on half adders, which add two bits and produce a sum and carry, and full adders, which also account for a carry added from a less-significant bit (pp. 239–242, *Digital Design with CPLD Applications and VHDL*). Full adders can be grouped together to make a parallel binary adder, with *n* full adders allowing two *n*-bit numbers to be added, generating an *n*-bit sum and a carry output (pp. 242–245, 247–252, *Digital Design with CPLD Applications and VHDL*).

A parallel adder can be converted to a two's complement adder/subtractor by including XOR functions on the inputs of one set of operand bits, say input B, allowing the operations $A + B$ or $A - B$ to be performed (pp. 253–255, *Digital Design with CPLD Applications and VHDL*). A control input, *SUB* (for SUBtract), causes the XORs to invert the B bits if HIGH, producing the one's complement of B. *SUB* will not invert B if LOW, transferring B to the parallel adder without modification. If *SUB* is also tied to the carry input of the parallel adder, the result is $(A + B + 0 = A + B)$ when $SUB = 0$ and $(A + \overline{B} + 1 = A - B)$ when $SUB = 1$, where $\overline{B}$ is the one's complement of B and $(\overline{B} + 1)$ is its two's complement.

Sign-bit overflow occurs when a two's complement sum or difference exceeds the permissible range of numbers for a given bit size (pp. 230–232, *Digital Design with CPLD Applications and VHDL*). This can be detected by an SOP circuit that compares the operand and result sign bits of a parallel adder, or by an XOR gate that compares carry into and out of the MSB (pp. 256–258, *Digital Design with CPLD Applications and VHDL*).

Using Components in VHDL

VHDL designs can be created using a hierarchy of design entities. For example, a parallel adder can be designed as the top level of a hierarchy that contains several VHDL full adder components. An advantage of this approach is that certain functions, such as full adders, seven-segment decoders, and so on, can be designed once and used many times in different design projects. The method of using components in a VHDL file is called **structural** design, in contrast to **dataflow** design, which uses signal assignment statements and similar constructs or **behavioral** design, which is based on descriptions of circuit operation.

Structural design requires:

1. One or more complete VHDL design files that can be used as components. These are separate from, but used by, the top level of the design hierarchy.
2. Component *declaration* in the top-level VHDL file, similar to the entity declaration of the component.
3. Component *instantiation* in the top-level VHDL file, which maps the inputs and outputs of the component to the ports and signals of the top-level design.

The general form of a top-level design entity using components is:

```
ENTITY entity_name IS
    PORT ( input and output definitions);
END entity_name;

ARCHITECTURE arch_name OF entity_name IS
    component declaration(s);
    signal declaration(s);
BEGIN
    Component instantiation(s);
    Other statements;
END arch_name;
```

The components in the above file are complete designs in their own right, defined in different files. As long as the folder containing the component is on a recognizable library path, the top-level file will compile, using the component files.

Device and pin assignments can be made directly in the VHDL files, without requiring the components to be inserted as symbols in a MAX+PLUS II graphic design file.

Structural design of parallel adders is discussed in more detail on pp. 247–252 of *Digital Design with CPLD Applications and VHDL.*

Procedure

Full Adders

1. The logic diagram for a full adder is shown in Figure 9.1. Use this diagram to write a VHDL file called ***drive:*\max2work\lab09\full_add.vhd.** Do not use the graphic editor. Set the project to the current file and save the design. Assign pin numbers to the design, as shown in Table 9.1. This can be done directly from the Assign menu in the MAX+PLUS II text editor.

Table 9.1 Pin Assignments for a Full Adder

Function	Device	Pin
a	SW1-1	34
b	SW1-2	33
c_in	SW1-3	36
c_out	LED1	44
sum	LED2	45

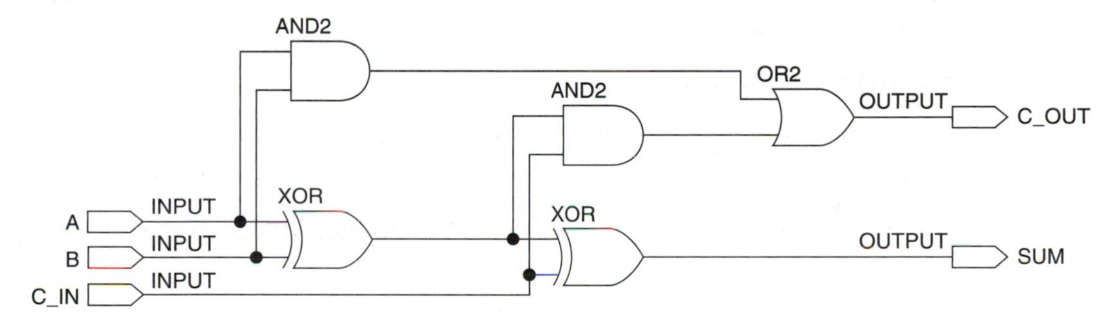

Figure 9.1 Full Adder

2. If you have not already done so, disable the CPLD outputs that are wired to unused LEDs. (E.g., for the Altera UP-1 board, include an output port called `unused : OUT STD_LOGIC_VECTOR(3 to 16)`, since only two out of 16 LEDs are used. Set the unused outputs HIGH, so that any LEDs connected to them will be off. The statement `unused <= (others => '1');` will accomplish this. For the RSR PLDT-2 board, the code for the unused outputs is the same, except for the statement `unused <= (others => '0');`)

   Assign pin numbers to the `unused` vector, corresponding to the standard connections for LED3–LED16 on the CPLD board (pins 46, 48, 49, 50, 51, 52, 80, 81, 4, 5, 6, 8, 9, and 10). Compile the file.

3. Create a simulation for the full adder. Run the simulation and show it to your instructor.

Instructor's Initials: ________

4. Download the full adder design to your CPLD board. (If you are using the Altera UP-1 board, be sure to invert the outputs of the full adder to make them active-LOW.)

   Take the truth table of the full adder to verify its operation. Show the result to your instructor.

Instructor's Initials: __________

Parallel Adder

1. Create an 8-bit parallel adder in VHDL, using a GENERATE statement and the component **full_add.vhd** from the previous section. Do not use a graphic design file. Permanently assign the carry input (C_0) to a logic LOW.

Note Make sure that the outputs of the full adder component are active-HIGH, even if you used active-LOW outputs in the previous section. Make sure that the sum and carry outputs of the parallel adder are active-LOW if you are using the Altera UP-1 board and active-HIGH for the RSR PLDT-2 board.

   Assign a device and compile the file.

2. Follow the examples shown below to add the unsigned binary numbers in Table 9.2, giving the sum in both binary and hexadecimal. Show the carry output separately.

   e.g.: 10111111 + 10000001 = 01000000 (Carry output = 1);
   Hexadecimal equivalent: BFH + 81H = 40H (Carry output = 1)

   e.g.: 00111111 + 00000001 = 01000000 (Carry output = 0);
   Hexadecimal equivalent: 3FH + 01H = 40H (Carry output = 0)

Table 9.2 Sample Sums for an 8-bit Parallel Adder

Binary Inputs	Carry	Binary Sum	Hex Equivalent
01111111 + 00000001			
11111111 + 00000001			
11000000 + 01000000			
11000000 + 10000000			

3. Use the sums calculated in Table 9.2 to create a simulation of the 8-bit parallel adder you created in procedure 1 of this section. The sums in the simulation must match those in Table 9.2. (You may have to account for active-LOW levels, depending on which CPLD board you are using.) Show the simulation to your instructor.

Instructor's Initials: __________

4. Assign pins to the 8-bit adder as shown in Table 9.3.

Table 9.3 Pin Assignments for an 8-bit Parallel Adder

Function	Device	Pin
a8	SW1-1	34
a7	SW1-2	33
a6	SW1-3	36
a5	SW1-4	35
a4	SW1-5	37
a3	SW1-6	40
a2	SW1-7	39
a1	SW1-8	41
b8	SW2-1	28
b7	SW2-2	29
b6	SW2-3	30
b5	SW2-4	31
b4	SW2-5	57
b3	SW2-6	55
b2	SW2-7	56
b1	SW2-8	54

Function	Device	Pin
c8	LED8	52
sum8	LED9	80
sum7	LED10	81
sum6	LED11	4
sum5	LED12	5
sum4	LED13	6
sum3	LED14	8
sum2	LED15	9
sum1	LED16	10
unused1	LED1	44
unused2	LED2	45
unused3	LED3	46
unused4	LED4	48
unused5	LED5	49
unused6	LED6	50
unused7	LED7	51

5. Compile the file and download the design for the 8-bit parallel adder to your CPLD board. Test the operation of the 8-bit parallel adder by applying the combinations of inputs *A* and *B* listed in Table 9.2. Show the results to your instructor.

Instructor's Initials: __________

Two's Complement Adder/Subtractor

1. Modify the 8-bit adder you created in the previous section to make an 8-bit two's complement adder/subtractor. Include an overflow detector that will turn on an LED when the output of the adder/subtractor overflows beyond the permissible range of values for an 8-bit signed number. Display the sum outputs on LEDs, as with the previous 8-bit adder, but also add a pair of seven-segment decoders to display the result numerically on the board's seven-segment display.

   Make the design entirely in VHDL, using components where required. Do not use a graphic file.

Notes:

- If you are using the Altera UP-1 board, recall that the LEDs and numerical displays are active-LOW. Binary sum outputs must be active-LOW to display the binary sum properly on the LEDs, but the internal signals connecting the adder/subtractor sum outputs to the seven-segment decoder inputs must be active-HIGH.
- If you are using the RSR PLDT-2 board, recall that the LEDs and numerical displays are active-HIGH. Internal signals for LED outputs and seven-segment decoder inputs can be the same, since they are both active-HIGH.
- If you are using a common-cathode seven-segment decoder (e.g., for the RSR PLDT-2 board), you will have to select some logic synthesis options in MAX+PLUS II to make the design fit into the EPM7128SLC84 CPLD. Select **Global Project Logic Synthesis** from the **Assign** menu (Figure 9.2). From the resulting dialog box, check the box for **Multi-Level Synthesis for MAX 5000/7000 Devices** (Figure 9.3).

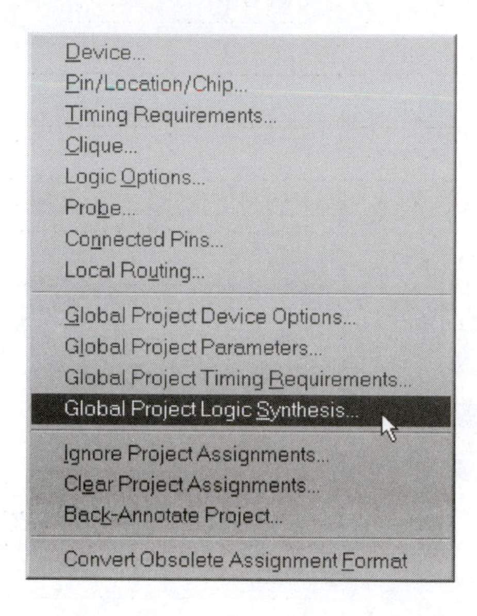

Figure 9.2
Selecting Synthesis Options (Assign Menu)

Figure 9.3
Selecting Multi-Level Synthesis

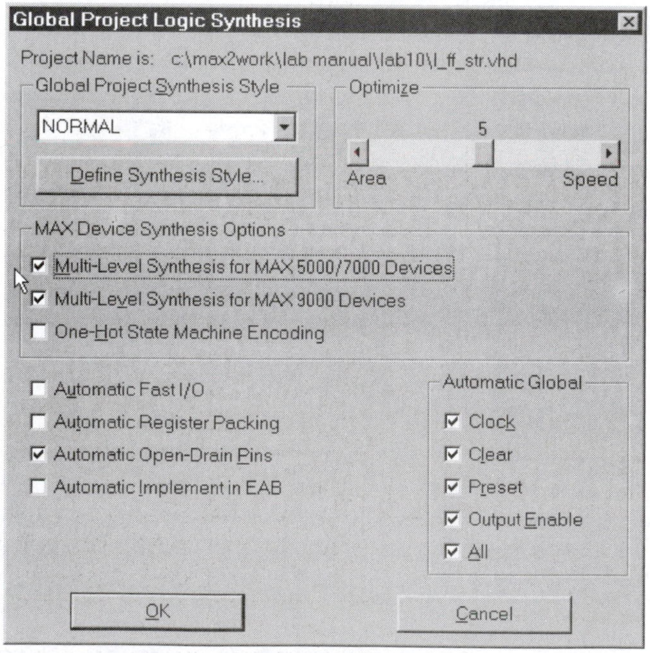

2. Follow the examples shown to add the signed binary numbers in Table 9.4, giving the sum in both binary and hexadecimal. Show the carry and overflow outputs separately from the sum. (Note that the carry output is not used for signed additions and subtractions. However, we retain the function so that the same circuit can be used for both signed and unsigned arithmetic.)

    e.g.: 01111111 + 01000001 = 11000000 (Overflow =1, Carry output = 0);
    Hexadecimal equivalent: 7FH + 41H = C0H

    e.g.: 00000000 (00000011 = 11111101 (Overflow = 0, Carry output = 0);
    Hexadecimal equivalent: 00H – 03H = FDH
    (two's complement: FDH = 11111101 = (-3_{10}))

Table 9.4 Sample Sums and Differences for an 8-bit Adder/Subtractor

Binary Inputs	Overflow	Carry	Binary Sum	Hex Equivalent	Two's Comp.
01111111 + 00000001					
11111111 + 00000001					
00000000 – 00000001					
00000000 – 01111111					
11000000 + 01000000					
11000000 + 10000000					

3. Leave pin assignments the same as for the 8-bit adder, except for the changes shown in Table 9.5.

Table 9.5 Pin Assignment Changes for 8-bit Adder/Subtractor

Function	Device	Pin
a1	Seven-Segment Digit 1	58
b1		60
c1		61
d1		63
e1		64
f1		65
g1		67
dp1		68
a2	Seven-Segment Digit 2	69
b2		70
c2		73
d2		74
e2		76
f2		75
g2		77
dp2		79
sub	MAX_PB1 (Altera UP-1) S5-1 (RSR PLDT-2)	11 12
overflow	LED7	51

4. Compile the file and download it to the CPLD board. Test the operation of the circuit by applying the *A* and *B* inputs from Table 9.4. Show the results to your instructor.

Instructor's Initials: ________

Assignment Questions

Complete problems 6.22, 6.30, and 6.34 in *Digital Design with CPLD Applications and VHDL.*

Lab 15

State Machines

Name ______________________ Class __________ Date __________

Objectives Upon completion of this laboratory exercise, you should be able to:

- Design, simulate and program a dual-sequence counter using VHDL.
- Design, simulate and program a traffic-light controller using VHDL.

Reference Dueck, Robert K., *Digital Design with CPLD Applications and VHDL*

Chapter 10: State Machine Design
10.2 State Machines With No Control Inputs
10.3 State Machines With Control Inputs
10.6 Traffic Light Controller

Equipment Required CPLD Trainer:
Altera UP-1 Circuit Board with ByteBlaster Download Cable, or
RSR PLDT-2 Circuit Board with Straight-Through Parallel Port Cable, or
Equivalent CPLD Trainer Board with Altera EPM7128S CPLD
MAX+PLUS II Student Edition Software
AC Adapter, minimum output: 7 VDC, 250 mA DC
Anti-static wrist strap
#22 solid-core wire
Wire strippers

Experimental Notes

A **state machine** is a synchronous sequential circuit whose states progress according to the inherent design of the machine and possibly according to the states of one or more control inputs. A synchronous counter is a simple state machine; it progresses in a fixed sequence which may vary with a control input, such as a directional control. Design of this type of state machine is described in detail in section 10.2 of *Digital Design with CPLD Applications and VHDL* (pp. 459–466). The Gray code counter described in this section can be combined with a 3-bit binary counter to make a circuit that will count in a binary sequence or a Gray code sequence, depending on the status of a control input.

A common state machine application is that of a traffic light controller. A state diagram and a description of the operation of this machine are given in section 10.6 of *Digital Design with CPLD Applications and VHDL* (pp. 490–492).

Procedure

Dual-Sequence Counter

1. The VHDL code in section 10.2 of *Digital Design with CPLD Applications and VHDL* describes a 3-bit counter that counts in Gray-code sequence. Modify the VHDL code to perform the following function:

Select a 3-bit binary or Gray code count, depending on the state of an input called **gray**. If **gray** = 1, count in Gray code. Otherwise count in binary.

Note VHDL only allows one clock evaluation per process (i.e., only one statement of the form `IF(clock'EVENT and clock= '1')THEN`). Therefore, the VHDL statements for the binary and Gray code counters must all be within the same IF statement that evaluates the clock. Each counter can be defined by its own CASE statement.

2. Create a simulation file for the dual-sequence counter, clearly showing the full Gray code count, binary count, and reset function. Show the simulation waveforms to your instructor.

Instructor's Initials: __________

3. Assign pin numbers to the outputs and disable any unused LEDs.
4. Compile and download this counter to a CPLD test board. Demonstrate the operation to your instructor.

Instructor's Initials: __________

Traffic Light Controller

A simple traffic light controller can be implemented by a state machine that has a state diagram such as the one shown in Figure 15.1.

The circuit has control over a North-South road and an East-West road. The North-South lights are controlled by outputs called **nsr, nsy,** and **nsg** (North-South red, yellow, green). The East-West road is controlled by similar outputs called **ewr, ewy, ewg.**

The cycle is controlled by an input called TIMER which controls the length of the two green-light cycles (s0 represents the EW green; s2 represents the NS green.) When TIMER = 1, a transition from s0 to s1 or from s2 to s3 is possible. This accompanies a change from green to yellow on the active road. The light on the other road stays red. An unconditional transition follows, changing the yellow light to red on one road and the red light to green on the other.

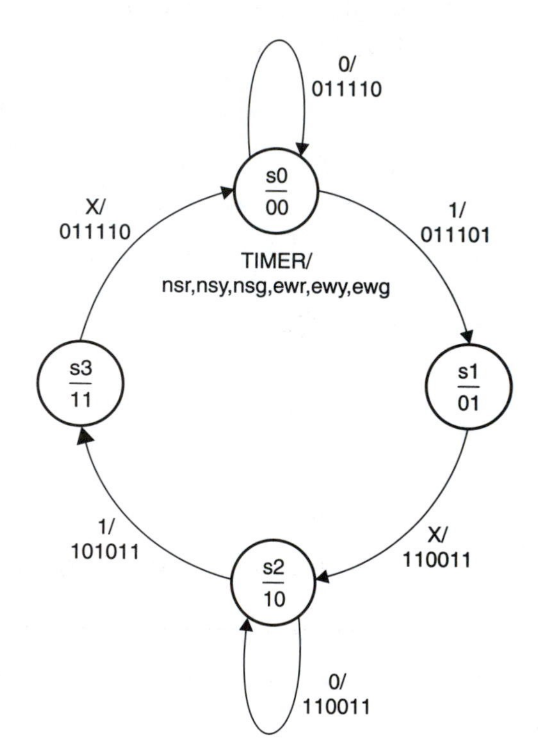

Figure 15.1 State Diagram for a Traffic Controller

The cycle can be set to any length by changing the signal on the TIMER input. (The yellow light will always be on for one clock pulse.) For ease of observation, we will use a cycle of ten clock pulses for any one road: 4 clocks GREEN, 1 clock YELLOW, 5 clocks RED. This can be generated by a mod-5 counter, as shown in Figure 15.2.

The clock divider brings the on-board oscillator frequency down to the range of visible observation for our CPLD board. A 25-bit counter is used for the Altera UP-1 board, which has an on-board oscillator frequency of 25.175 MHz. A 22-bit counter is

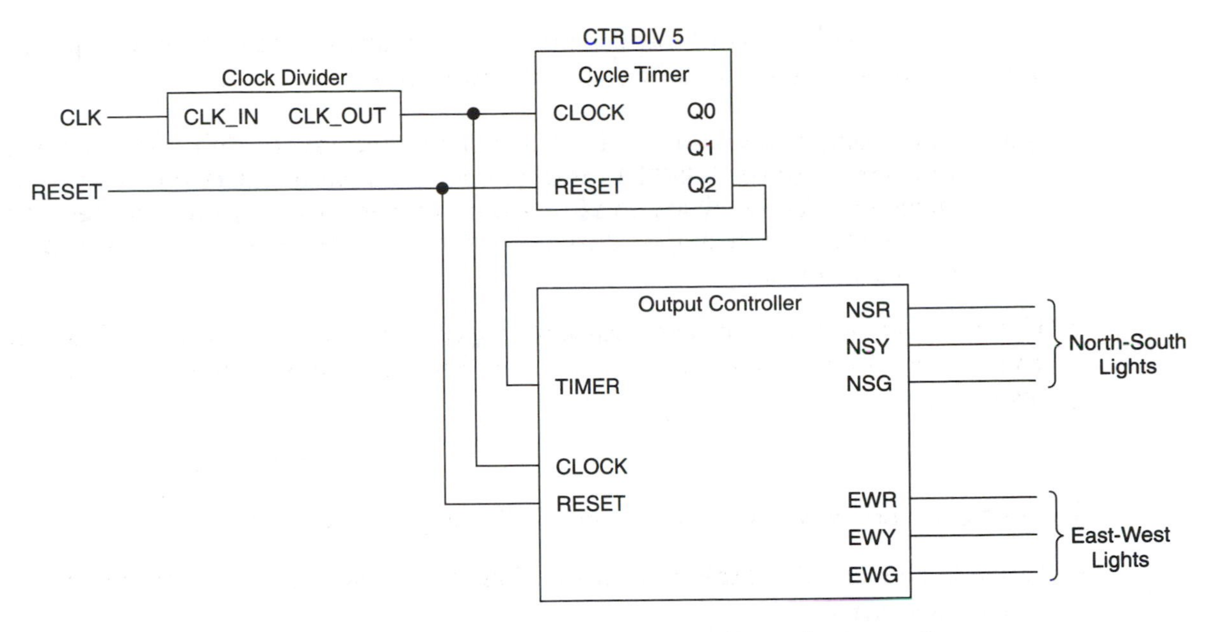

Figure 15.2 Logic Diagram for a Traffic Light Controller

suitable for the RSR PLDT-2 board, which has an on-board oscillator with a frequency of 4 MHz. Calculate the frequency of the state machine clock for your CPLD board.

$f =$ __________

Draw the timing diagram of the mod-5 counter in the space provided below:

CLK

Q0

Q1

Q2

How does the counter set the green-light cycle length to 4 clock pulses?

____

____

Creating a Traffic Light Controller in VHDL

1. Create a MAX+PLUS II file to implement the traffic controller state diagram shown in Figure 15.1, combined with the mod-5 cycle timer. The individual modules and the top-level of the hierarchy must all be done in VHDL.
2. Create a simulation that shows the combined operation of the output controller and cycle timer. Show the waveforms to your instructor.

Instructor's Initials: ________

3. Add a clock divider to the VHDL file for the traffic controller, as shown on the logic diagram of Figure 15.2. For the clock divider, use the module **clkdiv.vhd** and set its width to 22 (RSR PLDT-2) or 25 (Altera UP-1). Assign pins to the design so that the North-South lights correspond to LED1–LED3 on the CPLD board and the East-West lights correspond to LED9–LED11. Assign a pin to the state machine clock (so that it can be observed directly) and make it correspond to LED16. **Disable all other LEDs.** Make sure that the controller outputs are at the correct active level for the LEDs on your CPLD board.

4. Download the file to the CPLD board and demonstrate the operation to your instructor.

Instructor's Initials: __________

Traffic Controller with Walk Signal

1. Modify the VHDL files of the previous section to implement a traffic light controller with a walk signal, as shown in the logic diagram of Figure 15.3.

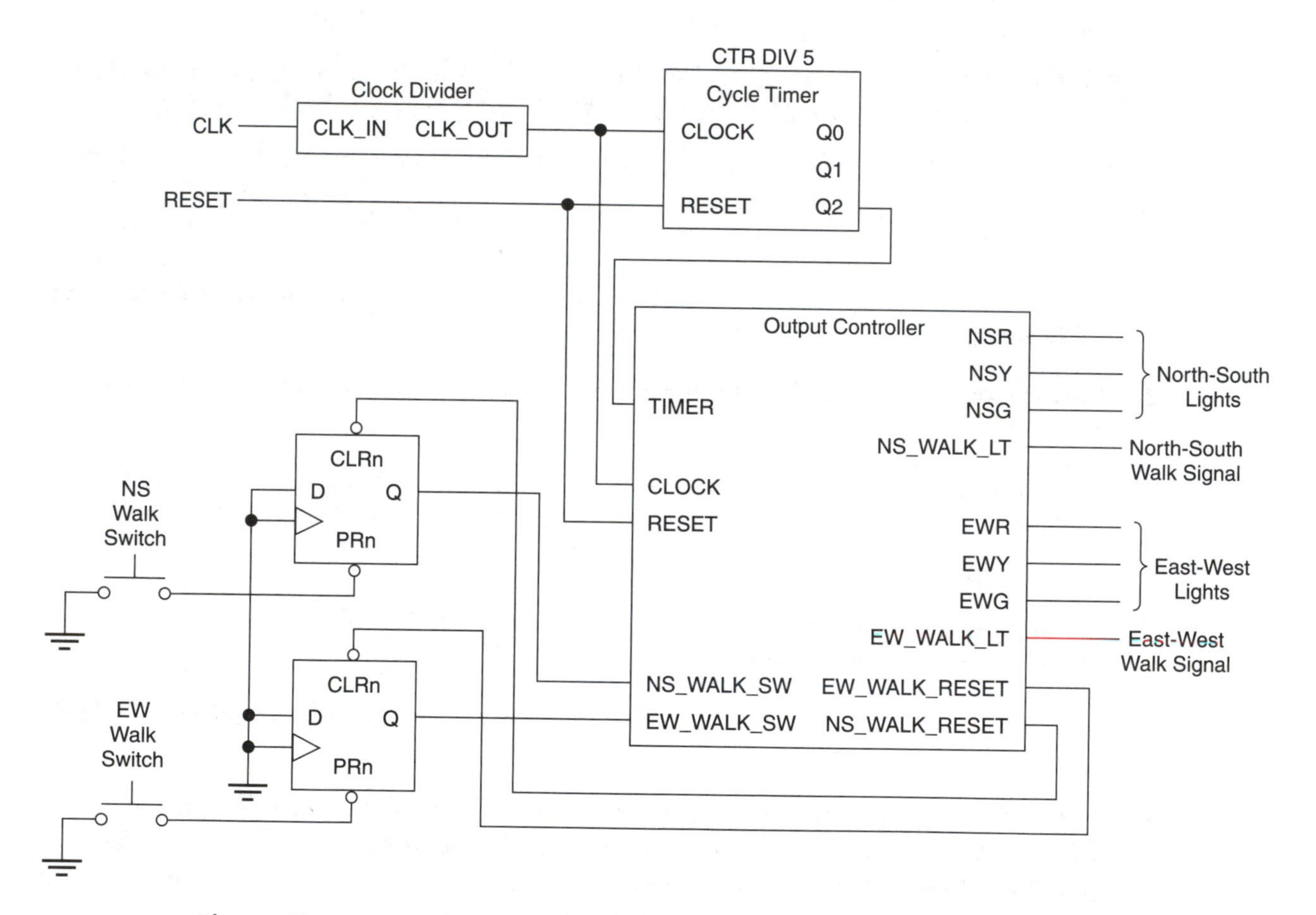

Figure 15.3 Logic diagram for a Traffic Light Controller with a Walk Signal

The walk signal goes on for one green cycle of the direction related to the switch. For example, when you press the NS switch, the North-South walk signal goes on for the next North-South green cycle. On the next NS green cycle, the walk signal is off unless the NS switch has been pressed again.

Lab 16

DAC-Based Function Generator

Name ______________________ Class __________ Date ______________

Objectives Upon completion of this laboratory exercise, you should be able to:

- Interface an integrated circuit Digital-to-Analog Converter to an EPM7128S CPLD on an Altera UP-1 or RSR PLDT-2 circuit board.
- Program an Altera CPLD to control the DAC for manual input.
- Program an Altera CPLD to make the DAC generate sawtooth, square, and triangle waves, with function selectable by DIP switch input.

Reference Dueck, Robert K., *Digital Design with CPLD Applications and VHDL*

Chapter 12: Interfacing Analog and Digital Circuits
12.1 Analog and Digital Signals
12.2 Digital-to-Analog Conversion

Equipment Required

CPLD Trainer:
Altera UP-1 Circuit Board with ByteBlaster Download Cable, or
RSR PLDT-2 Circuit Board with Straight-Through Parallel Port Cable, or
Equivalent CPLD Trainer Board with Altera EPM7128S CPLD
MAX+PLUS II Student Edition Software
AC Adapter, minimum output: 7 VDC, 250 mA DC
Anti-static wrist strap
Wire strippers
#22 solid-core wire
Solderless breadboard
±12-volt power supply
DAC0808 or MC1408 Digital-to-Analog Converter
High-speed op-amp
0.1 μF capacitor
0.75 pF capacitor
2.7 kΩ resistor
4.7 kΩ resistor
6.8 kΩ resistor
5 kΩ potentiometers (2)

Experimental Notes

A monolithic (one-chip) digital-to-analog converter (DAC), such as the DAC0808 or MC1408, can be easily interfaced to a CPLD for a static or dynamic output.

The CPLD can simply pass through the values of a set of DIP switches, or other digital inputs, so that the DC output can be adjusted to a desired value within its range. A procedure for calibrating such a circuit is detailed in *Digital Design with CPLD Applications and VHDL* (pp. 581–582).

The DAC can also be configured to generate a periodic time-varying output. To do so, the changing set of values is fed to the DAC inputs. If the output of a binary counter is connected to the DAC input, the result is a linearly-increasing output with an (ideally) instantaneous return to zero at the end of each cycle. This function is also known as a sawtooth or ramp waveform. An example of a DAC-based ramp generator is shown in example 12.7 (pp. 583–584) of *Digital Design with CPLD Applications and VHDL.*

Other waveforms can be generated by changing the output sequence of the counter that drives the DAC inputs. The output behavior of such a counter can be described by a series of VHDL statements. An example of a digital function generator based on stored values of DAC input is shown on pp. 639–641 of *Digital Design with CPLD Applications and VHDL.*

Procedure

DAC/CPLD Interface

1. Wire the circuit for a DAC0808 digital-to-analog converter onto a solderless breadboard, as shown in Figure 16.1. The pin numbers for the device are indicated in parentheses.

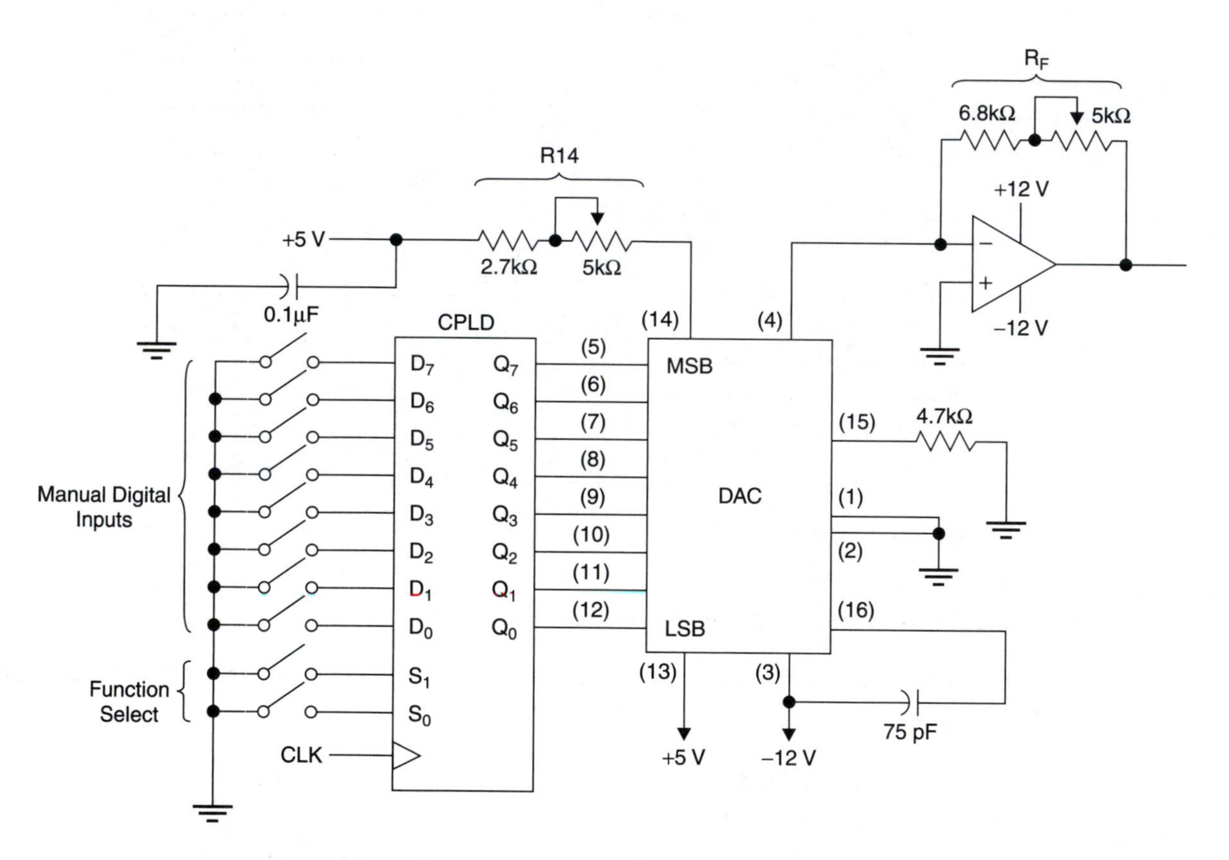

Figure 16.1 DAC Function Generator

2. Create a MAX+PLUS II file that will read the values of eight DIP switches at a set of CPLD input pins and pass them through the CPLD to a set of output pins without modification (i.e. an internal pin-to-pin connection). Although the DAC can be directly connected to the DIP switches, it is efficient to pass the signals through the CPLD so

that we can keep these same pin connections, without rewiring, for later parts of the lab.

3. Use the pin assignments shown in Table 16.1. Disconnect any LED connections for LED9 through LED16. Use the pins normally assigned to these LEDs in other projects to connect from the CPLD to the DAC inputs.

Table 16.1 Pin Assignments for DAC/CPLD Interface

Function	Device	CPLD Pin
D_7	SW1-1	34
D_6	SW1-2	33
D_5	SW1-3	36
D_4	SW1-4	35
D_3	SW1-5	37
D_2	SW1-6	40
D_1	SW1-7	39
D_0	SW1-8	41
Q_7	DAC pin 5 (MSB)	80
Q_6	DAC pin 6	81
Q_5	DAC pin 7	4
Q_4	DAC pin 8	5
Q_3	DAC pin 9	6
Q_2	DAC pin 10	8
Q_1	DAC pin11	9
Q_0	DAC pin 12 (LSB)	10
S_1	SW2-7	56
S_0	SW2-8	54
CLK		83

4. Compile and download the file to the CPLD board.

5. The resistor networks shown in Figure 16.2 allow us to set our input reference current and output gain to values within a specified range.

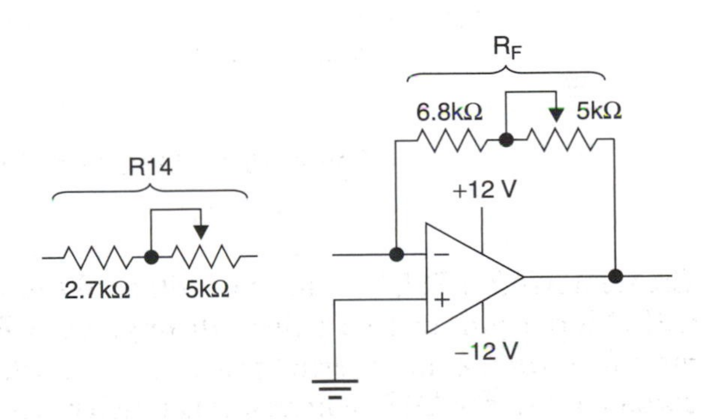

Figure 16.2 DAC Resistor Networks

Using the values shown in Figure 16.2, fill in Table 16.2 for the cases when V_o is at minimum and maximum, and when the pots are at their midpoint values. (R_{14} is the total value of the input reference resistance. R_F is the total resistance of the op-amp feedback network.) Assume the DAC input is set to 1111 1111. Show calculations in the space provided below.

Table 16.2 DAC Output Range

	R_{14} (Ω)	R_F (Ω)	I_{ref}(mA)	I_o(mA)	V_o(V)
Minimum V_o					
Maximum V_o					
Pots at midpoint					

Calculations:

6. Calibrate the DAC as follows:

Connect a Digital Multimeter (DMM) to the op-amp output. Apply power to the circuit and adjust the feedback pot for the minimum value of voltage. Set the input code to 10000000.

Adjust the R_{14} pot so that the output voltage of the op-amp is 3.4 volts. What value of I_{ref} does this correspond to? I_{ref} = ________.

Set the feedback pot so that the output voltage is 5 volts. What value of R_F does this correspond to? R_F = __________

Measure the output voltage of the DAC circuit for the digital input values in Table 16.3.

Table 16.3 DAC Output Voltages for Manual Inputs

Input Code	Output Voltage (Calculated)	Output Voltage (Measured)	Error (volts)	Error (LSB)
00000000				
00000001				
00000011				
00000111				
00001111				
10000000				
11000000				
11100000				
11111111				

State the calculated value of resolution (i.e., the value of 1 LSB) for this DAC.

Resolution = __________ volts

Instructor's Initials: __________

DAC Ramp Generator

1. Refer to the DAC-based ramp generator in example 12.7 (pp. 583–584) from *Digital Design with CPLD Applications and VHDL*. Create a circuit similar to the ramp generator in Figure 12.14 by programming an 8-bit counter into the EPM7128S CPLD. *The previous DAC interface should not be changed.* The counter driving the ramp generator should be clocked at about 1.57–2 MHz by the output of a clock divider. (The on-board oscillator of the Altera UP-1 board runs at 25.175 MHz. The RSR PLDT-2 board has an on-board oscillator with a frequency of 4 MHz.)
2. Assign pins to the counter design file so that the counter outputs are the same as the pins connected to the DAC inputs. Compile and download the MAX+PLUS II counter file.

3. Connect an oscilloscope to the DAC op-amp output. Draw the sawtooth waveform generated by the DAC. Measure its period and calculate the sawtooth frequency.

   T = __________; = f __________.

   Sketch of a sawtooth waveform:

4. Divide the counter clock frequency (f_c) by the DAC output frequency (f_{DAC}) to get an estimate of the number of clock pulses per sawtooth cycle. Since the clock signal from the on-board oscillator is divided by a clock divider inside the CPLD, the value of f_c cannot be measured directly unless the divided clock is brought out on a CPLD pin. Alternatively, you can measure the frequency at the LSB input of the DAC (pin 12) and multiply by 2.

   Compare the calculated ratio of f_c / f_{DAC} to the ideal value and determine the % error. Also state the % error of an oscilloscope measurement. (As long as you can measure to within the error of the oscilloscope, the measurement is reasonably accurate.) Estimate the oscilloscope error as follows:

   - Count the number of small divisions on the oscilloscope horizontal grid line.
   - Estimate what fraction of a small division it is possible to measure.
   - Divide the measurable fraction of a small division by the total number of small divisions and multiply by 100%.

   f_c / f_{DAC} = __________ clock cycles (measured)

   f_c / f_{DAC} = __________ clock cycles (ideal)

   %error (frequency measurement) = __________

   %error (oscilloscope screen) = __________

Instructor's Initials: __________

Other Functions

1. Program the EPM7128S CPLD to incorporate **sawtooth, triangle,** and **square** waveforms, as well as a **pass-through for manual DIP switch input** to the DAC. Program the CPLD using VHDL. The function driving the Q outputs will depend on the binary value of S_1S_0. Consider the information provided in Table 16.4.

Table 16.4 Criteria for Digital Waveforms

Waveform	Condition
Sawtooth	Binary-increasing input 00 to FF. Input then drops to 00 (can be generated by counter function applied to input)
Square	Input = 00 for first half-cycle; = FF for second half-cycle
Triangle	Input increases linearly to positive peak, then decreases linearly to negative peak (no sharp drop-off).

Use the input switch combinations shown in Table 16.5 for function select.

Table 16.5 Select Switch Values for Digital Waveforms

S_1	S_0	Function
0	0	Manual pass-through
0	1	Sawtooth
1	0	Square
1	1	Triangle

2. Connect an oscilloscope to the op-amp output and demonstrate the operation of the waveform generator.

Instructor's Initials: __________

CAUTION Do not disassemble the DAC circuit. It will be reused in Lab 17.

Assignment Questions

1. Answer the questions in the lab procedure.
2. Complete problems 12.18 and 12.20 in *Digital Design with CPLD Applications and VHDL.*

Lab 17

Analog-to-Digital Conversion

Name ______________________ Class __________ Date ____________

Objectives Upon completion of this laboratory exercise, you should be able to:

- Design, simulate, program, and test an interface between an EPM7128S CPLD and an ADC0808 Analog-to-Digital Converter.
- Determine the effect of sampling frequency on aliasing for an ADC0808 A/D converter.

Reference Dueck, Robert K., *Digital Design with CPLD Applications and VHDL*

Chapter 12: Interfacing Analog and Digital Circuits
12.3 Analog-to-Digital Conversion
12.4 Data Acquisition

Equipment Required

CPLD Trainer:
Altera UP-1 Circuit Board with ByteBlaster Download Cable, or
RSR PLDT-2 Circuit Board with Straight-Through Parallel Port Cable, or
Equivalent CPLD Trainer Board with Altera EPM7128S CPLD
MAX+PLUS II Student Edition Software
AC Adapter, minimum output: 7 VDC, 250 mA DC
Anti-static wrist strap
#22 solid-core wire
Wire strippers
Solderless breadboard
±12-volt power supply
Analog Function Generator (sine wave output)
ADC0808 Analog-to-Digital Converter
DAC0808 or MC1408 Digital-to-Analog Converter
High-speed op-amp
0.1 μF capacitor
0.75 pF capacitor
2.7 kΩ resistor
4.7 kΩ resistor
6.8 kΩ resistor
10 kΩ resistors (3)
5 kΩ potentiometers (2)
1N4004 diodes or equivalent (4)

Experimental Notes

The ADC0808 Analog-to-Digital Converter (ADC), shown in Figure 17.1, is a successive approximation ADC with eight multiplexed inputs. The ADC converts one channel at a time, as selected by the binary combination of inputs ADDA, ADDB, and ADDC, where ADDC is the most significant bit. The ADC requires a pulse on the address latch enable

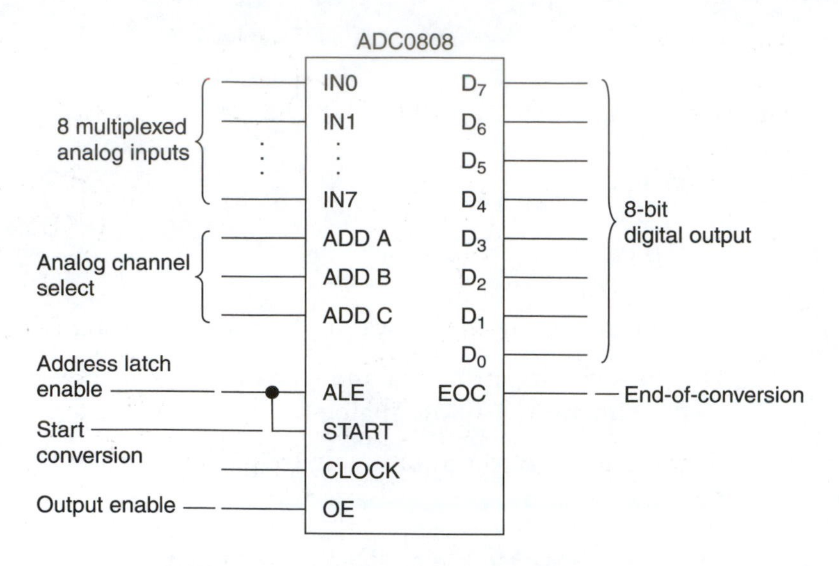

Figure 17.1 ADC Analog-to-Digital Convertor

(ALE) line and on the START input to begin the conversion process. An ADC output called EOC (End of Conversion) goes LOW near the beginning of the conversion process and then goes HIGH to indicate that the conversion is complete. Figure 17.2 shows the relative timing of the device. Further details are given on pp. 605–610 in *Digital Design with CPLD Applications and VHDL.*

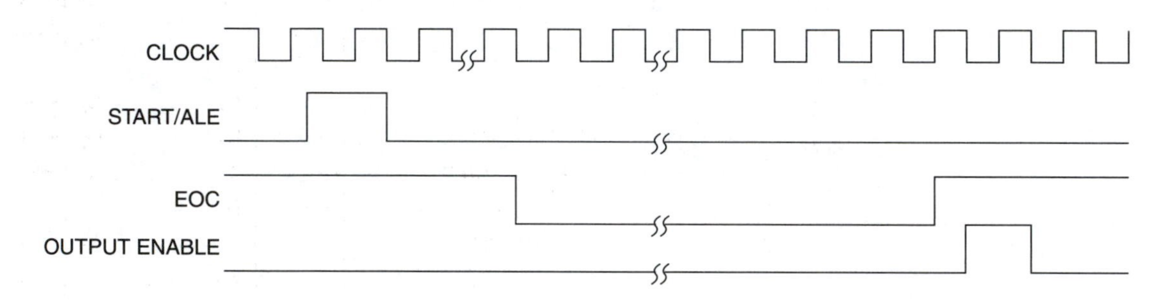

Figure 17.2 ADC0808 Timing

Procedure

CPLD-to-ADC Interface

1. Write a VHDL file for an ADC0808 converter, based on the state diagram of Figure 17.3 on page 163. Create a MAX+PLUS II simulation for the controller. Show the controller waveforms to your instructor.

Instructor's Initials: ________

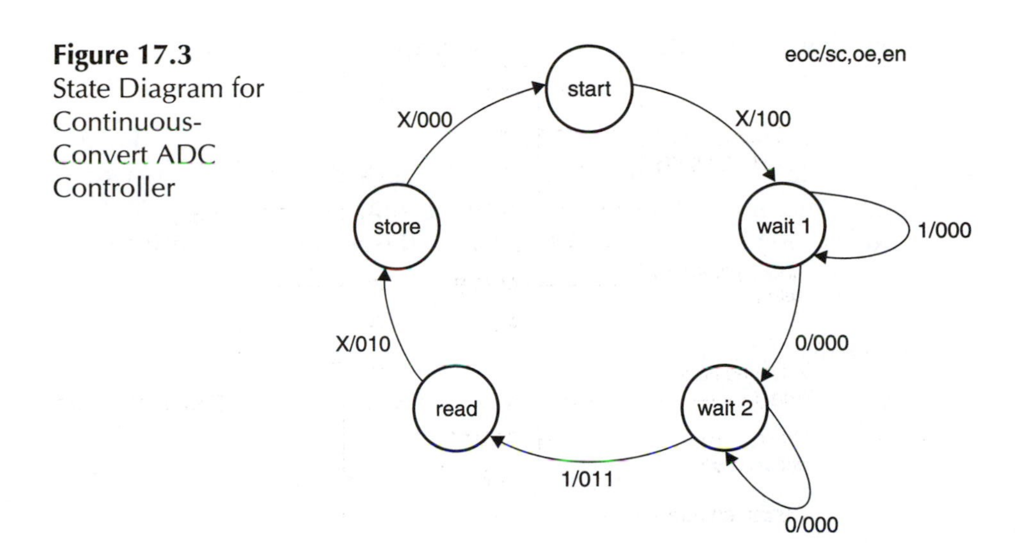

Figure 17.3
State Diagram for Continuous-Convert ADC Controller

2. Add a clock divider and an output latch to your VHDL controller, as shown in Figure 17.4. The clock frequency of the ADC0808 must be kept below 1280 kHz. Divide the clock of your CPLD board accordingly. (Altera UP-1 board: on-board oscillator is 25.175 MHz; RSR PLDT-2: on-board oscillator is 4 MHz.) Add a pair of seven-segment decoders to the latch outputs (Q_7 through Q_0).

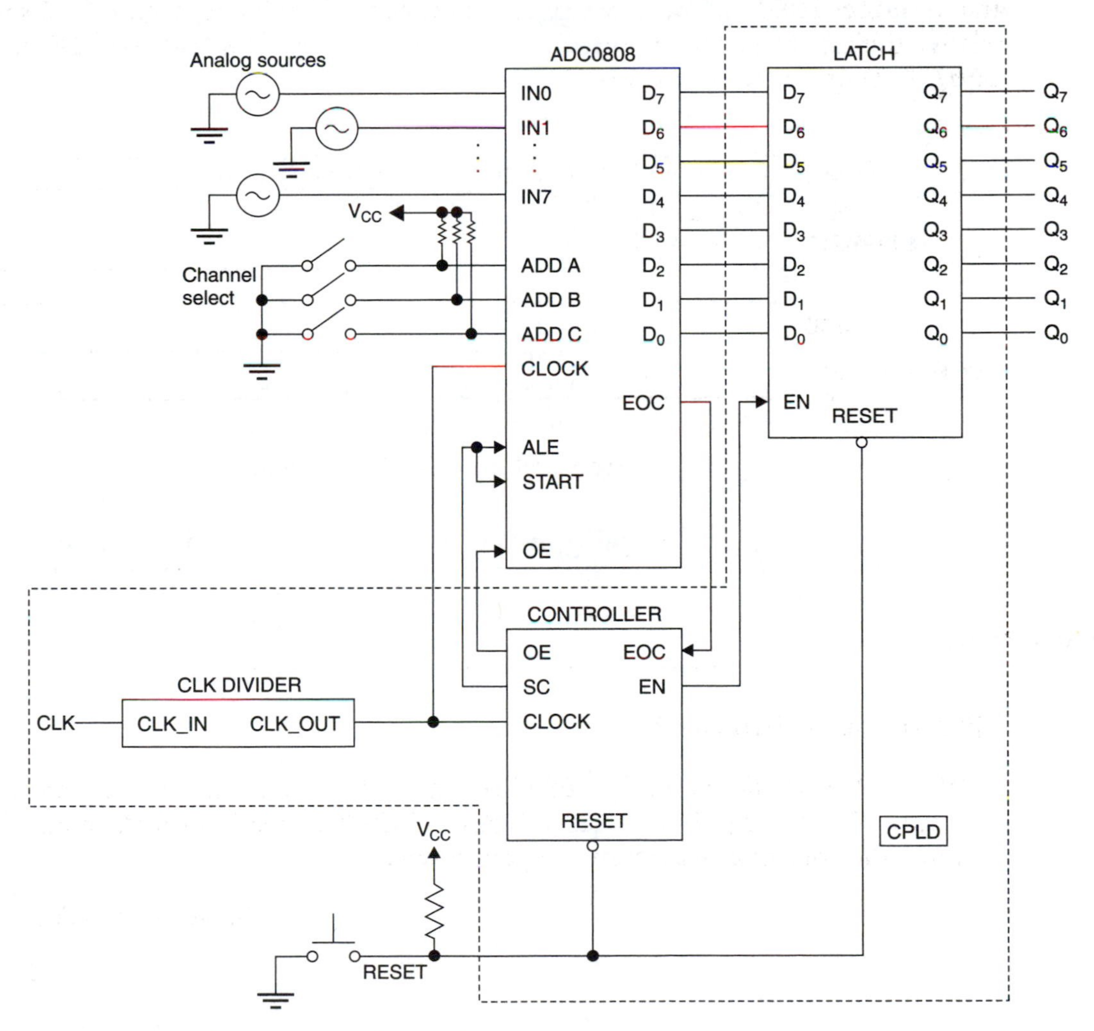

Figure 17.4 ADC Interface with One Output Channel and Manual Input Channel Selection

3. Assign pins to the CPLD for the design shown in Figure 17.4. Be sure to include pin assignments for the seven-segment displays. Compile and download the file to your CPLD board.

4. Refer to the data sheet for the ADC0808 (available on the internet from National Semiconductor at **<http://www.national.com>**). Connect the ADC0808 to the CPLD board, as shown in Figure 17.4. Connect the current and voltage protection networks shown in Figure 17.5 to input channels IN0 and IN1.

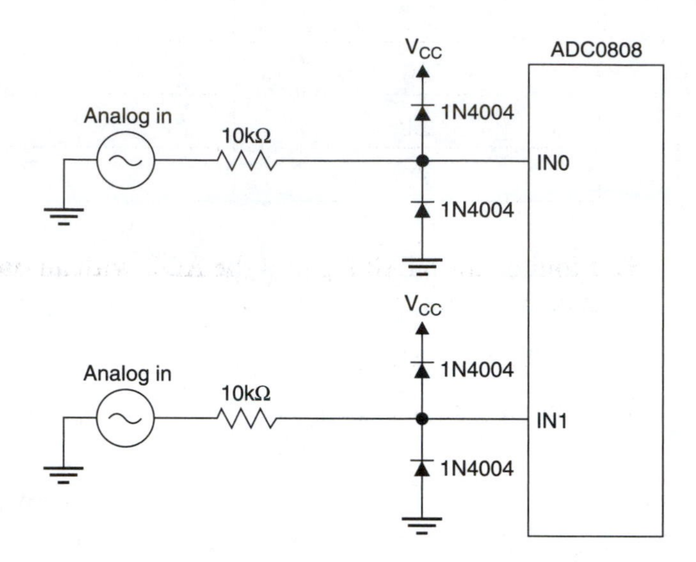

Figure 17.5 Current and Voltage Protection for ADC Inputs

5. Connect a variable 5-volt power supply to IN0 and select the channel by the position of the address DIP switches. Before turning on the power, show the wiring of your circuit to your instructor.

Instructor's Initials: ________

6. Turn on the power and vary the analog inputs. Fill in Table 17.1 with measured and calculated values for the ADC0808. For each entry, measure the input voltage with a digital multimeter (DMM) and note the output on the CPLD board's seven-segment displays.

Table 17.1 Analog-to-Digital Converter Test Data (Channel 0)

Digital Output	Measured Analog Input	Calculated Analog Input
00		
01		
02		
80		
C0		
FF		

7. Repeat the measurements on Channel 1 and fill in Table 17.2.

Table 17.2 Analog-to-Digital Converter Test Data (Channel 1)

Digital Output	Measured Analog Input	Calculated Analog Input
00		
01		
02		
80		
C0		
FF		

8. Monitor the START pin of the ADC with an oscilloscope. What is the time between successive conversions, based on the spacing of pulses on the START line?

____

9. Demonstrate the operation of the circuit to your instructor.

Instructor's Initials: __________

Time-Varying Waveforms and Aliasing

1. Modify the VHDL code for your ADC controller system so that it no longer has the seven-segment outputs. Connect the ADC outputs to the bipolar DAC circuit shown in Figure 17.6.

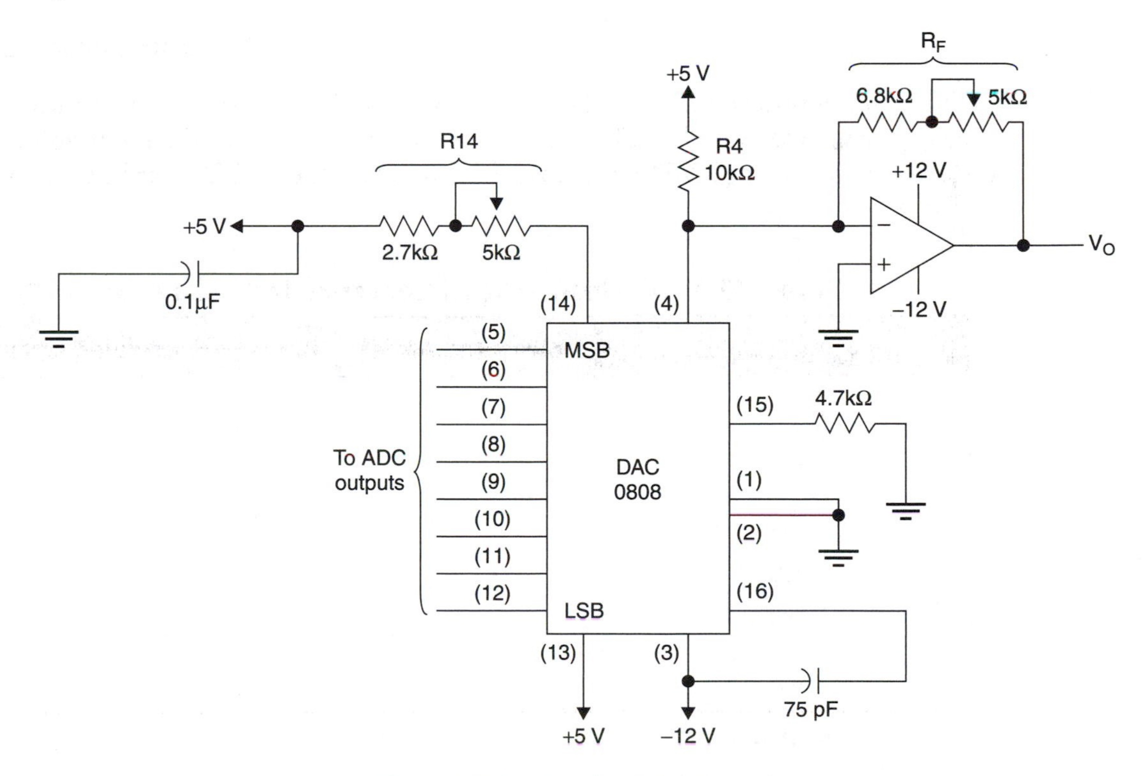

Figure 17.6 Bipolar DAC Circuit

2. Adjust the sine wave output of an analog function generator so that its outputs vary from 0 volts to +5 volts. Start with the generator frequency at 100 Hz. Connect the generator to an analog input of the ADC and select the appropriate channel. Monitor the output of the DAC to see the reproduced analog voltage.

3. Increase the frequency of the analog generator until you noticeably see an alias frequency appearing. What is the highest analog frequency that can be applied to the circuit without developing an alias frequency component? ________

   How does this compare to your calculation of the maximum frequency that can be sampled with this system. Show calculations below. (See examples 12.13 and 12.14 (pages 610–611) in *Digital Design with CPLD Applications and VHDL.*)

   Calculations:

4. Demonstrate the operation of the circuit to your instructor.

Instructor's Initials: ________

Assignment Questions

1. Answer the questions in the lab procedures.

2. Complete problems 12.38 and 12.44 from *Digital Design with CPLD Applications and VHDL.*